与曾叔叔闲聊 2 C

Science Chats with Uncle Reggie
与曾叔叔聊科学
新生儿科教授的洞见

曾振锚著 Reginald Tsang

与曾叔叔聊科学
新生儿科教授的洞见

与曾叔叔闲聊 2

与曾叔叔聊科学——新生儿科教授的洞见

作者／曾振锚
中文编辑／郑名芝
设计排版／Busybees Design Consultants Ltd.

版次／二零一九年七月初版
　　　二零二四年四月初版五刷
©2019 曾振锚

国际书号：978-1-5456-7190-0

版权所有　请勿翻印
本书中文经文取自圣经——新标点和合本，版权为香港圣经公会所有，承蒙允许使用

Copyright © 2019 by Reginald Tsang

Coffee with Uncle Reggie 2
Science Chats with Uncle Reggie

by Reginald Tsang
Chinese Editor/ Sofia Cheng
Design/ Busybees Design Consultants Ltd.
Edition/ First edition, July 2019
　　　　 First edition (Fifth printing), April 2024

ISBN 978-1-5456-7190-0

All rights reserved solely by the author. The author guarantees all contents are original and do not infringe upon the legal rights of any other person or work. No part of this book may be reproduced in any form without the permission of the author.

Scripture quotations taken from the New International Version (NIV). Copyright © 1973, 1978, 1984, 2011 by Biblica, Inc.™. Used by permission. All rights reserved.

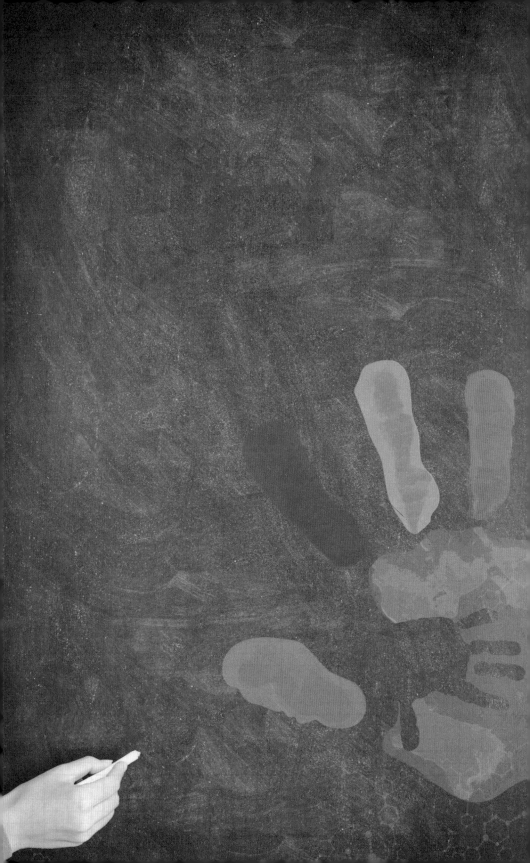

目录
Table of Contents

序言 ... 6
Foreword

引言及致谢 ... 22
Introduction and Acknowledgements

1. 逻辑与设计：我们是否合乎逻辑？ ... 28
Logic and Design: How Logical are We?

2. 海滨漫步：寓言 ... 34
Just Walking on the Beach: An Allegory

3. 汉考克和贝聿铭的精彩签名 ... 38
Great Signatures: John Hancock and I. M. Pei

4. 上海细胞 ... 44
Shanghai Cell

5. 世上最好的疼痛医生 ... 48
The Best Pain Doctor in the World

6. 恭喜您，您的小宝宝有生理突变？ ... 58
Congratulations, Your Baby Has a Mutation?

7. 或许你会得良性突变呢！ ... 64
You Might Get a Good Mutation!?

8. 达尔文的眼睛：荒谬至极 ... 70
Darwin's Eye: Absurd in the Highest Degree

9. 但它们和我们这么像！ ... 80
But They Look So Much Like Us!

10. 欢腾雀跃的猿妈妈 ... 88
Jubilant Mama Ape Story

11. 阑尾不是附属品 96
An Appendix is not an Appendix

12. 我的扁桃体是残留的？ 104
My Tonsils are Vestigial?

13. 鲑鱼情结 112
I Love Salmon

14. 天才就是天才：天才中的天才 124
Genius is Genius: The Genius of Genius

15. 语言的奥秘（上）：语言令我着迷 136
Mystery of Language Part A: My Fascination with Language

16. 语言的奥秘（下）：语言从何而来？ 148
Mystery of Language Part B: Where Does Language Come From?

17. 做个怀疑者（上）：信条恒变 160
Be a Skeptic Part A: Dogmas Come and Go

18. 做个怀疑者（下）：基因、骨头和皇帝 174
Be a Skeptic Part B: Genes, Bones, and Emperors

19. 何谓红十字？ 184
What's the Red Cross?

20. 个人的见证：临床医生科学家的观点 190
Personal Testimony: A Clinician Scientist Perspective

后记 200
Postscript

作者简介 206
About the Author

Foreword

Wonderful to know that the second "Coffee with Uncle Reggie" book is about to be published. I have thoroughly enjoyed reading the first book, and have always looked forward to reading articles that come out regularly on Reggietales.org, some of which are included in this second book. I have the privilege of knowing Reggie for over five decades, as both of us graduated from the same medical school and we practiced in the same medical specialty prior to our retirement. Reggie's knowledge and wit were legendary. These attributes have made the two "Coffee with Uncle Reggie" books so enjoyable and entertaining to read. Although the articles are mainly "stories" from Reggie's personal experiences and life journey, they help open our mind to soul-searching insights into our own lives and important perspectives on the world that we live in, which once grasped becomes life-changing. They impart great wisdom that can transform our lives to give us joy and peace and fulfillment in a world full of darkness, fears and disappointments. I highly recommend *Science Chats with Uncle Reggie*, not only for those who like us come from the intellectual world of academic medicine, but for everyone who might be finding life somewhat confusing and meaningless.

The Reverend Professor **Victor Yu**
Emeritus Professor, Faculty of Medicine,
Monash University, Melbourne, Australia

序言

很高兴知道"与曾叔叔闲聊"系列的第二册即将出版。我非常喜欢第一册,并常常期待阅读 Reggietales.org 上定期发表的文章,其中一些文章会收藏在第二册中。我有幸认识振锚超过 50 年,我们在同一所医学院毕业,在退休前又执业于相同的医疗专科。振锚知识广博,又风趣,是众所周知的。这两点使他的"与曾叔叔闲聊"系列能够让读者感到愉快和有趣。虽然这些文章主要是来自振锚的个人经历和人生旅程的"故事",但是它们能帮助我们反思,探索我们对于自己生命和周遭世界的重要观念,一旦掌握,便成为改变生命的力量。这些睿智故事,能够改变我们的生命,在这个充满黑暗、恐惧和失望的世界中给予我们喜乐、平安和满足。我极力推荐《与曾叔叔聊科学——新生儿科教授的洞见》,它不仅适合那些和我们一样来自医疗学术世界的人,也适合每一位对人生感到困惑和缺乏意义的读者。

余宇熙教授
澳洲墨尔本蒙纳士大学医学院荣誉教授

与曾叔叔聊科学
新生儿科教授的洞见

With wit, wisdom and winsomeness, Uncle Reggie shares stories that will touch and perhaps transform your heart. A great read!

David Stevens, MD, M.A. (Ethics),
Chief Executive Officer, Christian Medical & Dental Associations, USA

Heartiest congratulations to *Science Chats with Uncle Reggie* of the "Coffee with Uncle Reggie" series. It is a great honor to write this foreword for a remarkable effort of a compilation of anecdotes and tales that dear Uncle Reggie has put together. Part I has been an inspiration to so many who have read it, including myself, and I believe this new sequel will continue to be so if not more. I am most honored also to have had the opportunity to be visiting scientist at Cincinnati where I got to know Uncle Reggie. This book I believe may serve to spiritually nourish the soul just as we need to physically feed our bodies. I wish to draw attention to readers who may not know that Uncle Reggie is an icon and expert in nutritional support of the preterm infant. In his self-declared "third retirement", Uncle Reggie successfully complemented professional with his wealth of life experiences to help "nourish" the soul of others by illustrating that the heavenly conduct is actually very tangible in our daily lives, in very down-to-earth selected readings. What a great tribute to God and a magnificent venture this is, indeed. I sincerely hope it will succeed to draw many to "see" that there is a Higher being who controls our lives and this universe. Nothing occurs by chance, as in my opinion the complexity of life and the live systems are too intricate and delicate to be explained in full. This also, I surmised the more in depth I explored while conducting my PhD research, and as such, I

风趣、有智慧又吸引人,曾叔叔分享的故事不但会触动你的心,也许更会改变它。这是一本很好的书!

大卫·史蒂文斯医生,**文学硕士(伦理)**
美国基督徒医师及牙医协会首席执行官

衷心祝贺"与曾叔叔闲聊"系列的《与曾叔叔聊科学——新生儿科教授的洞见》出版。曾叔叔费尽心血将他的轶事和故事结集成书,我非常荣幸能够为这本书写序言。许多读过第一册的读者(包括我自己)都深受启发。我相信这第二册会继续启发更多读者。我很荣幸有机会在辛辛那提做访问学者,在那里我认识了曾叔叔。我相信这本书会在精神上滋养我们的心灵,正如我们要喂养我们的身体一样。我特别希望读者留意到,曾叔叔是早产儿营养学专家,是这方面的权威。在他自称"第三次退休"的时期,曾叔叔以其专业,加上丰富的生活经验,透过一些朴实的文章,说明上天的作为在日常生活中其实显而易见,从而"滋养"他人的心灵。这确实是向创造者致敬,也是伟大的创举。我真诚地希望,这本书能够成功吸引许多人"看到"有一位更高的创造者,掌管着我们的生活和这个宇宙。在我看来,没有任何事情是出于偶然的,生活和生命的系统错综复杂,我们未能解释通透。我猜我做博士研究时探索得越深入,就越能认同此书第 20 章和后记所写的,并发现它引人入胜之处。令人遗憾的是,越来越多人主张要相信和依靠自己才会成

can personally identify with Chapters 20 and postscript in this book, and find truly endearing. Regrettably, there appears to be an increasing propaganda in the calling to believe and depend on oneself as humanly possible, and not to trust another to succeed. As a result when society is faced with adversity, people become more cynical, and hope is trivialized. In fact, hope in our walk with God and viewing life perspectives through His lenses may just be the solution in living a rich, long, successful life as the stories in *Science Chats with Uncle Reggie* reassuringly and lovingly convey, albeit loudly and clearly. I commend this wonderful contribution wholeheartedly and am sure *Science Chats with Uncle Reggie* will be of tremendous success, perhaps spawning yet another sequel in the future.

Fook-Choe Cheah, MD, PhD
Professor and Chair of Paediatrics
UKM Medical Centre, Kuala Lumpur, Malaysia
7th January, 2019

Congratulations on your *Science Chats with Uncle Reggie*. It's so wonderful for me to read that book.

Ran Namgung, MD
Professor of Pediatrics, Yonsei University College of Medicine
Former President of Korean Society of Neonatology

Dr. Reggie Tsang is a world-renowned physician-scientist, a prolific writer on science and societal themes of utmost importance to mankind and, perhaps most importantly, he is a mentor to generations of physicians and scientists who have made their marks and have become leaders in their respective fields. He led the division of neonatology at the prestigious Cincinnati Children's Hospital and

功，而不是信任他人。因此，当社会面临逆境时，人们便变得更愤世嫉俗，没有盼望。其实，《与曾叔叔聊科学》中的故事，响亮清晰，却又亲切明确地传达一个信息：我们要心存盼望与上天同行，并以他的眼光去看待生命，这样我们就能够活得丰盛、长久而成功。我衷心赞扬这些精彩的文章，并确信《与曾叔叔聊科学》将取得重大的成功，将来或再出版续集。

Fook-Choe Cheah 医生，博士
马来西亚吉隆坡 UKM 医学中心
儿科教授兼主席
2019 年 1 月 7 日

恭喜你出版《与曾叔叔聊科学——新生儿科教授的洞见》。能够读这本书，真是太棒了。

Ran Namgung 医生
延世大学医学院儿科教授
韩国新生儿科学会前主席

曾振锚医生是世界知名的医生科学家，也是一位多产作家，他的著作谈及科学和社会等对于人类至关重要的主题。更重要的是，他是一代又一代医生和科学家的导师，这些医生和科学家表现卓越，在各自的专业范畴成为领袖。曾医生曾在著名的辛辛那提儿童医院医学中心领导新生儿科 14 年。在他的

Medical Center for 14 years. Under his tutelage, Cincinnati Children's Hospital and University of Cincinnati established the first perinatal research institute in the United States of America and the world's "capital" when it comes to neonatal nutrition, especially calcium and vitamin D metabolism. His research efforts resulted in continuous funding for a 15-year span by the National Institute of Health (NIH). This culminated in over 400 peer-reviewed publications, and the classic textbook on neonatal nutrition, *Nutrition During Infancy*. In spite of all these academic accomplishments, Dr. Tsang's real interest is mostly in training next generation of neonatologists. He had helped shaped the career of countless neonatologists and nutrition researchers all over the world. He always preaches the true meaning of living. He is a role model extraordinaire! Dr. Tsang's book series, "Coffee with Uncle Reggie" and articles in Dr. Tsang's website, Reggietales.org, are poignant reminders of what is important in life. They are a "must read" if for people seeking the true meaning of life. We shall forever be grateful for the impact he has made in our lives.

Henry T. Akinbi, MD
Professor of Pediatrics
Cincinnati Children's Hospital Medical Center and The Department of Pediatrics,
University of Cincinnati College of Medicine
Cincinnati, Ohio, USA

Forewords from Coffee with Uncle Reggie

It has been said that if you walk with the wise in your life, it is your honor, and you should be grateful to have met such a person in your life.

带领下,辛辛那提儿童医院和辛辛那提大学联手成立了美国第一间围产期研究所,成为新生儿营养学(特别是钙质和维生素D代谢)的世界"首府"。他的研究工作获得美国国家卫生研究院(NIH)15年的持续拨款,最终发表了400多篇经由同行评审的论文,以及有关新生儿营养的经典教科书《婴儿营养原理与实践》。尽管取得了这些学术成就,曾医生的真正兴趣主要在于培训下一代的新生儿科医生。他协助过世界各地无数新生儿科医生和研究营养学的科学家规划他们的事业。他总是传扬生命的真谛。他是非凡的榜样!曾医生的著作"与曾叔叔闲聊"系列以及网站 Reggietales.org 上的文章,提醒我们什么是生命中重要的事情。对于寻求生命真谛的人来说,这些是"必读"的作品。我们将永远感激他对我们生命的影响。

Henry T. Akinbi 医生
美国俄亥俄州辛辛那提市辛辛那提大学医学院
辛辛那提儿童医院医学中心儿科系儿科教授

《与曾叔叔闲聊》序言

有人曾说,如果你在人生中与智慧之人相伴前行,是你的荣幸。在人生中遇到这样的人是值得感恩的。

曾振锚医生是我在1994年8月在华西大学"偶遇"的智慧之人。他不仅有智慧,而且拥有大爱大德。自此,我们在儿

I had a "chance encounter" with Dr. Reggie Tsang in August 1994 at West China University: he was just such a wise person, who had great love and benevolence. From that time onwards, we were partners striding together, unceasingly, in the pediatric world, and striving to set up the bridge for communication and interchange of China-American pediatric scholars. With great love and patience, Dr. Tsang personally dedicated his strengths and knowledge to the areas that needed him.

I always remember his gentle smile, but I do not remember any expression of anxiety. In any tense or anxious moments, he would smile away the problem. I remember his Mandarin was like those of Chinese ethnic origin who grew up in America, but his Putonghua was kind and gentle. He tried very hard to communicate clearly, always in a friendly tone.

I remember his physical hardship when he often traveled among many cities of China from Cincinnati, USA. It seems that at any time, he did not stop thinking and planning how to bring the best of American academics to China. And how to create the conditions for young physicians and nurses to have opportunities to learn in the USA. I know that he tried also to help meet their needs and make their stay in Cincinnati a very pleasant one, so that they could focus their energies on study, instead of being too homesick.

I remember he was always uniquely carrying a bag that had his pillow inside, so that he could protect and support his back prevent pain... Just remember his very long travel distances of more than 100-thousand mile journeys, how did he manage to complete all those air travels? Through all his strenuous efforts, he established many partnership ventures between hospitals and medical schools in China and Cincinnati Children's Hospital Medical Center. He was the emissary

科的世界一直结伴前行，务求架起中美儿科学者沟通交流的桥梁。他更以博大的爱心和恒心，身体力行地在需要他的地方奉献自己的力量和知识。

我总记得他的微笑，但记不起他曾有任何焦虑的表现。即使紧张或者焦虑，他也用微笑化解。记得他说普通话的时候，是带着在美国长大的华人的语调，但是他的普通话亲切而温柔。他总是尽量表达得清晰，语气和善。

记得他频频往返于中国各个城市和美国辛辛那提市的艰辛。他似乎任何时候都从来没有停止思考，计划着如何把美国最好的学术研究带到中国；如何创造条件，让年轻的医生和护士获得赴美学习的机会。我知道他更为身处辛辛那提的中国医生和护士提供帮助，让他们安心，让思乡之情变成学习动力。

记得那只他总是装在挎包里的枕头。坐着时，他的腰要靠这只枕头一直垫着才不会痛……再想着长达超过十万八千里的旅途，他是怎样坐着完成飞行旅程的？在他的努力下，中国多间医院和医学院与辛辛那提儿童医院医学中心开展了合作的项目。他是中美儿科医学合作协同发展的使者。

不，他就是天使。

记得 2004 年，他的夫人身患重病，然而在照顾夫人的日子里，他也始终想着我们。他在美国主编的《早产儿营养：基础与实践指南》是一部早产儿临床护理的经典。当他听说中国没有这样的专业书的时候，就立即同意我将它翻译成中文，于 2009 年在中国出版。他花了一番功夫，免去了此书的版税和

for development of cooperative agreements between China and American pediatrics.

No, indeed he is an angel.

I remember that in 2004, his wife had a serious illness, yet while he was looking after his wife, he was always thinking about us also. In America, he edited the classic "Bible" for preterm infant clinical care, *Nutrition of the Preterm Infant: Scientific Basic and Practical Guidelines*. When he realized that this kind of specialized book was not available in China, he immediately agreed that I could translate the book into Chinese, so that we could publish the book in 2009. Under his efforts he waived any charges and copyright issues, so that Chinese pediatricians could be benefited.

I have met his wife and daughter, and they both beautiful and kind from their deepest nature. Without saying a word, they touch your spirit. Truly, I'm very honored to have met and known Dr. Reggie, and I am very grateful.

In all the stories that Dr. Reggie has written for you in this book, it's good that you read them. They are also genuine and true. I believe that, just like me, you would be full of gratitude.

Meng, MAO, MD, Professor of Pediatrics;
President, West China Second University Hospital, Sichuan University 2001-2010;
President, Chengdu Women's and Children's Central Hospital 2011-2014;
Group Leader of Child Health Care, China Pediatrics Society of
China Medical Association 2009-present

Reggie was my senior by eleven years at the Medical School, The University of Hong Kong. By the time when I was still half way through my Pediatric training, he was already an internationally renowned neonatologist and guru in neonatal nutritional research.

其他收费,让中国儿科医生因此受益。

  我见过他的夫人和他的女儿。她们从内而外洋溢出来的美丽与善良,不需言语,便可以打动你的心灵。真的,能与振锚医生相遇、相识,我很庆幸,也深怀感激。

  振锚医生在这本书为你记录的这些故事,你读就对了。一切都是那么自然而真切。相信你一定会与我一样,充满感恩。

毛萌医生,儿科学教授
四川大学华西第二医院院长(2001-2010)
成都市妇女儿童中心医院院长(2010-2014)
中华医学会儿科学分会儿童保健学组组长(2009-现在)

  振锚是比我早 11 年毕业的香港大学医学院师兄。当我还在玛丽医院当实习儿科医生期间,早年同样在玛丽医院接受儿科培训的振锚,已经是国际知名的新生儿科专家和新生儿营养

Nicknamed "Mr. Calcium" in the Pediatric world for his supremacy in calcium and vitamin D research, he was also well known to us as one of the most distinguished alumni of the University, and a role model for Pediatric trainees to follow. It was not surprising that when Reggie took early retirement from clinical practice in 1994 to devote his time to the works of MSI, the news created a shockwave of disappointment and disbelief through the Pediatric community in Hong Kong. As things turned out, Reggie's selfless and bold decision actually brought him closer to us. In the course of his many service trips to China, he often had to stop over Hong Kong, and would look us up when he did. During these short visits he always preferred staying in the small local MSI office rather than checking into a hotel. "Just to save a few dimes for the organization", so he said. Personal comfort never seemed to be his concern or else he would not have made so many difficult journeys to the rural and poverty-stricken South-western China. In those days, merely travelling to those places was physically very demanding, not to mention the hardship he had to endure living in accommodations that often lacked even the most basic daily amenities.

Having travelled numerous miles in rural China and brought education to tens of thousands of poor village children, Reggie has now "retired" again but only to take up a new project of sharing his life experience with the rest of the world. The first volume of his story book *Coffee with Uncle Reggie* is a great collection of real stories he gathered from his many travels in Asia and China. They are stories of Adventure, with a wealth of words of wisdom and spiritual lessons that are inspirational to all, young and old. I am most delighted that Reggie would tell more of his stories this time in a bilingual book. Written in

研究的权威。振锚对钙质和维生素 D 的研究有杰出的成就，令他在儿科界中有"钙先生"的昵称。他是香港大学最杰出的校友之一，也是儿科医学生的榜样。难怪振锚在 1994 年提早退休、投入国际医疗服务机构（MSI）的工作时，香港儿科界对这震撼的消息既感失望，又觉得难以置信。事实证明，振锚无私而大胆的决定实际上使他和我们更接近。他频繁地前往中国提供医疗服务，经常都要停留香港，这时就会来找我们。短暂停留香港时，他总是喜欢住在 MSI 的小办公室，而不是入住酒店。他说："要为机构省一点经费。"个人舒适与否，他从来不担心，否则他不会千辛万苦地前往落后而贫困的中国西南地区。当年要到这些地方，单是长途跋涉已经非常艰辛，还要忍受连最基本日常设施都缺乏的住宿环境，肯定是吃了不少苦头。

振锚的足迹踏遍不少中国农村，并为数以万计贫穷的农村小孩子带来教育的机会。他现在已经再次"退休"，却又以新的方式与世界各地的人分享他的人生经验。他第一册的《与曾叔叔闲聊》，收集了他多次在亚洲和中国旅程中的真实故事。这些冒险故事充满了智慧之言和启示人生的教训，不论长者或年青人都会觉得鼓舞人心。我非常高兴振锚这次以双语讲述自己的故事。以英文写作、附中文翻译的《与曾叔叔闲聊》，更

English with Chinese translation, *Coffee with Uncle Reggie* would be able to reach people all over China, people who Reggie cares so much about.

Tai-fai Fok, Pro-Vice Chancellor and Vice President,
Choh-Ming Lee Professor of Paediatrics, The Chinese University of Hong Kong

Serving others is at the heart of everything Christian. If you think serving others for a higher calling is drudgery and something to avoid, please think again. Through Dr. Reggie Tsang's many engaging vignettes you will find the Lord working through a man's life to fill it with many global friends and meaningful kingdom work. There is adventure, like eating fried bees and hanging onto a patient who fell over the fifth floor railing of a hospital. As a neonatologist, Dr. Reggie cared for the physical lives of pre-born infants, but he was also an effective "doctor of the hearts and souls" of untold children and youth, and adults, around the world. Read and be inspired to invest your life in the Lord's work just as Uncle Reggie did.

Bruce Chester, Chairman of the Board, Medical Services International, USA;
Dedicated to medical missions in China

Coffee with Uncle Reggie contains a charming collection of stories, words skillfully woven together, not just to tell a tale, but more importantly to reflect special lessons that Uncle Reggie has learned along the way. While on this journey, he has been able to interpret his circumstances and communicate them in a way that demonstrates the goodness and greatness of the Creator and the intricacies of His creation. These tales will bring a smile to your face!

David Leung, MD;
President, MSI Professional Services, Hong Kong;
Physician, Evergreen China, Shanxi

能够接触到全中国的读者，也就是振锚非常关心的人。

霍泰辉教授
香港中文大学副校长
卓敏儿科讲座教授

服务他人是基督徒生活的核心。如果你觉得为回应一个更高的呼召而服务他人是一件苦差和要避免的事情，请再想一想。通过曾振锚医生的许多动人小故事，你会发现主如何使用一个人的生命，让他结交世界各地的朋友，并做有意义的工作。当中有冒险的时候，例如吃油炸的蜜蜂，和拉住一个要跨过医院五楼栏杆跳下去的病人。作为一位新生儿科医生，曾医生关心母腹中胎儿的身体健康，同时也是世界各地无数儿童、青少年以及成年人的"心灵医生"。请阅读这本书，你会受到启发，将自己的生命投入主的工作中，就像曾叔叔一样。

布鲁斯·切斯特
美国国际医疗服务机构委员会主席
致力从事中国的医疗服务工作

《**与曾叔叔闲聊**》系列是一连串引人入胜的故事，以文字巧妙地编织在一起，不仅仅是讲故事，更重要的是反映出曾叔叔在过程中学习到的特别教训。在这些经历中，他在理解他的处境和演绎故事的时候，更能够把造物主的善良和伟大，以及其创造之复杂精密表现出来。这些故事会使你会心微笑！

梁启予医生
香港国际医疗服务机构专业服务总裁
山西永青服务中心医生

Introduction and Acknowledgements

Welcome to this book, *Science Chats with Uncle Reggie*, in the series, Coffee with Uncle Reggie. I'm writing the book in the same "style," as *Coffee with Uncle Reggie*, as a chat with you while drinking coffee, in Cincinnati, Seattle, China or Thailand. Just picture me talking to you *face to face*. I have done that, countless times, with young and old. Especially in a Cincinnati café chain called Panera; one time I was even there 3 times in one day. The cafe should have a "frequent flyer" system for me! I wish you fun and joy as we explore the various angles I am taking, much like how I view my life experiences.

Life is truly a journey and I often try to learn from the real-life experiences of others, and especially their stories. This is particularly a major reason why I wrote this book, to introduce a perspective that has inspired me all my life, and led me into many exciting and meaningful experiences that I hope I can share with you. Feel free to share your stories to me also, since I just love stories, at rctsang@gmail.com or WeChat uncleReggie.

For me, I have always wished to apply my strong interest in medicine and science to be helpful to others through my medical training, medical teaching, and "3 retirements." For example, more than 20 years ago, I took my "first early retirement" to help start medical missions to rural China. More than 10 years ago, I took my "second retirement" to work with youth and missions, and helped to welcome numerous medical scholars

引言及致谢

欢迎来到"与曾叔叔闲聊"系列的《与曾叔叔聊科学——新生儿科教授的洞见》。这本书延续《与曾叔叔闲聊》一书的写作风格,我和你就像在辛辛那提、西雅图、中国或泰国一起喝咖啡聊天,你不妨想像一下这个情景。我已经无数次这样与青年人和成年人面对面交谈。特别是在辛辛那提 Panera 咖啡连锁店,我曾经一天三次到那里与人聊天。这家店应该给我"常客"的待遇!让我们来一起探索我看事情的各种角度,分享我如何看待我的生活经历。我希望在这个过程中你们感到有趣和快乐。

人生的确是一个旅程,我经常尝试从别人的真实生活经验中学习,特别是发生在他们身上的故事。这正正是我写这本书的主要原因——介绍启发了我一生的观点,这观点带给我许多令人兴奋和有意义的经历,希望能与你们分享。也请你通过 rctsang@gmail.com 或微信 uncleReggie 与我分享你的故事,因为我实在是非常喜欢故事。

对于我来说,一直希望用自己对医学和科学的兴趣做些对人有益的事,包括医学培训、医学教学和我的"三次退休"。例如二十多年前,我"第一次提早退休",去协助开展中国农村的医疗服务。十多年前,我"第二次退休",致力于青年人服务,并帮助我们的辛辛那提儿童医院迎接了众多医学学者。

to our Cincinnati Children's Hospital. Now I have started into my "third retirement" as story writer.

My love of medicine and science motivated *my scientific research*, to discover the amazing mysteries of design in the human body. My medical career tied together science, faith and medicine, *as a mission* from a loving God. Because of all this, I have a passion to somehow communicate this great blessing in my life to those I can meet and chat with.

For decades, I taught children and youth; college, postgraduate and medical students; and medical faculty; in America, Latin America, Asia, and Europe; in at least 150 cities. Because of my passion for science and faith issues, I naturally gravitate towards this theme often. I particularly enjoyed giving such lectures to many medical scholars that came to Cincinnati for training. But the greatest fun was the chance to chat with them informally during "English corner," at our Children's Hospital cafeteria, where scholars *got to chat* with native English speakers. Or at other informal activities, such as dinners and scenic tours of the lovely city of Cincinnati, which I loved to lead. The theme of faith and science invariably and naturally came up often during these encounters, especially when we saw the beauty of nature, and visited the beautiful churches and cathedrals of Cincinnati.

There are many people to thank for this book, because they have inspired me in different ways, to explain science and faith concepts in a better way. With time and experience, I trust that my communications have gotten more precise and concise, and hence have led to this work. I wish, especially to thank the great team involved in this production, the CVSG (Cincinnati Visiting Scholars Group) team lead by Peter Yang, Amy Zhao, Mary Fan, John Bascom and Mary Anne Lucas; and the Media Team of Felicity Tao, Jenny Liao, Yang Dixia, Eileen Mok, Bill Chan, Paul

现在我进入了"第三次退休",成为故事作家。

我对于医学和科学的热爱激励着**我的科学研究**,去发现人体设计的神奇奥秘。我的医学事业把科学、信念和医学联系在一起,是**天意使然**。因为这一切,我热衷于把我生命中这个伟大的祝福传达给与我相遇和聊天的人。

近几十年来,我教过孩子和青少年,大学生、研究生和医学生,还有医学教授;曾到过至少 150 个城市,足迹遍布美国、拉丁美洲、亚洲和欧洲。由于我对科学和信念方面的热忱,我经常谈论这个主题。我特别喜欢给来到辛辛那提进修的许多医学学者提供这种讲座,但是最大的乐趣是在我们儿童医院餐厅的"英语角"非正式地与他们聊天,他们在那里有机会与英语为母语的人**闲聊**。或者在其他非正式活动中,例如共进晚餐,并由我带队观赏辛辛那提市的美妙景色。在这些聚会中,特别是欣赏美丽的自然风景和参观辛辛那提美丽的大教堂的时候,几乎一定自自然然会谈到信念与科学的话题。

非常感谢很多人对这本书的支持。他们以不同的方式启发了我,使我得以用更好的方式解释科学和信仰的概念。凭借时间和经验,我相信我的沟通方式变得更加精确和简洁,从而促成了本书。我希望特别感谢参与本书成书的优秀团队 CVSG(辛辛那提访问学者团),这个团队由杨礼华、Amy Zhao、Mary Fan、约翰·巴斯科姆和玛丽·安娜·卢斯卡带领。还要感谢包括陶晴、Jenny Liao、杨迪霞、Eileen Mok、陈启源、容振威和 Richard Kwong 的媒体团队,他们通过各种方式传播曾叔叔的

Yung, and Richard Kwong. My deep appreciation for their wonderful faithful work in disseminating the Uncle Reggie stories through my Reggietales.org website, Facebook, YouTube, WeChat, my *Coffee with Uncle Reggie* book, and now the printing of this book.

I had previously been shy of publicly thanking my dear wife, for reasons of modesty, but now that I am 78 years of age, I should *throw modesty to the winds*, and acknowledge the 50 years plus of faithful support and encouragement, and especially her tolerance for my hiding in writing this book! And, as we should remember in all our life, all thanks be to God, the Father of us all, the Origin of all, and especially of science and faith!

Finally, over the last few years, I have come to deeply appreciate the meticulous work of Matthew Fung, ML Kwan, Sofia Cheng, Wendy Chu, and the graphics design team, without which this book would not have appeared.

Onwards, as I love to say!

Uncle Reggie / Reginald Tsang MD, Professor Emeritus of Pediatrics, Cincinnati Children's Hospital, Former Director of Neonatology, Founder and Executive Director Perinatal Research Institute, and Vice Chairman Department of Pediatrics, Cincinnati, USA. Currently quietly residing in Newcastle, Seattle, USA.

故事，包括通过我的 Reggietales.org 网站、脸书、YouTube、微信，和《与曾叔叔闲聊》一书，到现在这本书的印刷发行。我深深地感谢他们出色忠实的工作。

我以前一直不好意思公开感谢我亲爱的妻子，是因为谦虚，但是现在我已经 78 岁了，我应该**毫无顾虑地**感谢她五十多年来的忠诚支持和鼓励，特别是她容忍我藏起来写这本书！还有，正如我们一生都应该记住的，我们要感谢上天，我们所有人的父亲，万物的起源，特别是科学和医学的起源！

最后，深深感谢冯肇熙、关文龙、郑名芝、朱韵怡，以及平面设计团队，没有他们近年一丝不苟的工作，这本书将无法呈现给大家。

正如我常常说，继续前进吧！

曾叔叔 / 曾振锚医生
美国辛辛那提儿童医院儿科荣誉退休教授
前新生儿科主任
围产期研究所创办人及前执行总监
前儿科系副主席
目前静居于美国西雅图的纽卡斯尔

翻译：Grace Lee

1. Logic and Design: How Logical are We?

I have been personally involved for decades in writing medical grant proposals, receiving grant awards, advising others on grant writing, and reviewing proposals, and I can safely say that probably I have been involved in more than 1,000 grant proposals. And what is glaringly obvious is that grants hinge critically on excellent *logic and design*, without which the grant proposal basically fails.

The medical grant proposer has to present his or her *logic and design lucidly*, which will reflect whether the study itself is viable, and whether the proposer can execute it. And what is even more important, but often forgotten, is the ultimate basic *unspoken assumption*, that *the body itself has to be totally logical and perfectly designed*, for the proposal to work. It cannot be functioning "randomly" or "by chance."

For example if, at the end of the study, the result is contrary to our hypothesis, we know immediately that it is *not* that the system is wrong, but that *our study concepts* are wrong, and we have to rethink the investigation. We cannot blame the system. The system is always logical and well designed; it is *our thinking* that might not. Without a doubt, if the human body has no logic and no design, any research on it would be impossible.

It is quite clear that in the body, anything, everything, anywhere, and everywhere embodies this perfect logic and design. It is *inescapable*. The system is infinitely complex, invisibly complex, instinctively complex,

1. 逻辑与设计：我们是否合乎逻辑？

几十年来，我一直亲身参与医学基金提案的撰写，获得基金资助，建议其他人撰写提案，并且帮助他们修改申请书，保守的估计，我已经参与了超过 1,000 项基金申请。显而易见的是，优秀的**逻辑和设计**是提案获得接纳的关键，否则那份提案基本上是失败的。

医学基金申请者必须**清晰地**写出他或她的**逻辑和设计**，这将反映出研究课题本身是否可行，以及申请者是否能够将提案付诸实行。更重要但是我们又常常忘记的，是最基本的**潜在的前设**，就是人体本身必须完全合乎逻辑并有完美设计，提案才会是可行的。它不能只是在"随机"或"偶然"的情况下运行。

举例来说，如果在一项医学研究结束后，研究结果与我们最初预计的结果相反，我们立刻知道这并不是人体系统的错误，而是**我们的研究观念的错误**，我们必须重新思考调查。我们不能责怪人体系统。人体系统必定是很有逻辑而且设计得很好，只是**我们的想法**可能不对。毫无疑问，如果人的身体没有逻辑、没有设计，那么任何对于人体的研究都是不可能的。

很明显，人体的一切，任何部位、所有部位，都体现了这种完美的逻辑和设计。这是**无可否认**的。人体系统无限地复杂、

impossibly mathematically complex. All these super complexities point to a super intelligent and logical design, and thus a super intelligent and logical Designer.

In a great article in *Newsweek* on "How Human Life Begins," many years ago, the author goes into great detail about how complexity pervades the system, and how beautiful the whole system is, so much that the author claims that the entire complexity is *"as beautiful as the Sistine Chapel!"* The widely acknowledged most wonderful artwork of Michelangelo *is the art work* in the Sistine Chapel. Whether the author meant this or not, the center of the Chapel actually depicts the *classic account of creation by God of man.* God is shown dramatically with outstretched hand reaching out to man, to create him and to make contact with him, in the *origin of origins* account.

生命的起源如同西斯廷小教堂，教堂的主要特色是米开朗基罗在教堂穹顶所画的《创世记》。

The beginning of life is like the Sistine Chapel: the main feature of that chapel is Michelangelo's Creation account.

1. 逻辑与设计：我们是否合乎逻辑？
1. Logic and Design: How Logical are We?

无形地复杂、本能地复杂、难以置信的精确地复杂。这一切都超级复杂，都指向一个超级有智慧和逻辑的设计，从而指向一位超级有智慧和逻辑的设计师。

多年前在《新闻周刊》上有一篇很好的文章，关于"人类生命的起源"。文章作者详细描述人体系统每个细节是多么复杂，整个系统是多么绚丽，他甚至宣布，这一切如此复杂，"**如同西斯廷小教堂一样华美！**"大家公认，米开朗基罗最精彩的艺术杰作就是在西斯廷小教堂**穹顶的画作**。无论作者是否这个意思，小教堂的中心确实描绘了**人类的上主创造世界的经典场面**。在这**最初之起源**的场面中，上主戏剧性地伸出手，伸向人类，去创造他，接触他。

这位作者对于自然界之精致和人类起源之复杂极为欣赏，以至她在那篇热情洋溢的文章里赞叹："**似乎**是个奇迹！"然而，对于我来说，这篇文章最有趣的是编辑在编辑栏里改写了这个观点，写道："这**就是**奇迹！"可能很多人都不会注意到这一点，但是这"似乎"是个奇迹和这"就是"奇迹有天壤之别，而所有的研究发现都指向后者：完美的逻辑和设计无处不在，这真是个神奇的发现。

翻译：Daisy Wang

The author is so impressed by the sophistication in nature and how human life could begin with such complexity, that she exclaims in her enthusiastic article, "*seems like* a miracle!" However, for me, I thought that the most interesting thing of this article was that the editor rephrased the sentiment, in the editor's box, as "*it's* a miracle!" Many might not even notice this, but the difference between it "seems like" a miracle and "it's" a miracle is literally astronomical, and all findings point to the latter: perfect logic and design is pervasive, a truly miraculous finding.

1. 逻辑与设计：我们是否合乎逻辑？
1. Logic and Design: How Logical are We?

"似乎是个奇迹"对比"这就是奇迹"……是哪一个呢？
"Seems like a miracle" versus "it's a miracle"...which is it?

2. Just Walking on the Beach: An Allegory

As I'm walking on a white sands beach of the Pacific coast, the rain begins to fall. I spot a blue rain speckled rock among many others. Obviously many thousands of years of sea water have washed over and over its surface, polishing and polishing it till it now feels smooth and glazed with its original rock color, whether green, yellow, blue or brown. It's all very naturally charming, and I enjoy picking up the prettiest pebbles on different beaches, to put them on my desk at home, to remind me of the beauty of nature.

I walk further and pick up a bowl. It is sculpted well and seems to be derived from a coconut shell that has been polished and made for use in a kitchen, most likely by a long ago native. I think I can actually use it; it looks like it would hold a hot evening soup quite well without leaking. And I love collecting native things.

I walk further and I pick up a jade carving of a small tree with bright green leaves and sharply pink flowers. With a bit of washing and rinsing off of the bits and pieces of water plants entangled among the leaves, I could show this artwork in my living room. I conclude that this jade carving must be the work of a skilled Chinese craftsman, maybe even very recently.

Suddenly, I'm surprised to see a very large building along the beach front, and so I'm curious, and want to check it out. I enter its ornate

2. 海滨漫步：寓言

当我漫步在太平洋岸边白色的沙滩上，雨开始飘落。在无数石头当中，我注意到一块带有蓝色雨点纹的石。显然，千万年来海水一遍一遍冲刷它，打磨它，使它现在摸起来是如此光滑，呈现着石头本身绿色、黄色、蓝色、棕色的色彩。这一切都是大自然的魅力。每到不同的海滩，我都喜欢捡拾漂亮的鹅卵石，把它们摆放在家里的书桌上，提醒我大自然的美丽。

继续漫步，我又捡起一个碗。这个碗雕刻得不错，看来是个经过抛光的椰子壳，是个厨房用具，很可能是很久以前本地土人做的吧。我想我应该用得着这个碗，用它盛晚餐的热汤应该蛮不错，似乎也不会漏水。而且我喜爱收集有当地特色的物品。

继续漫步，我捡起一座玉雕小树，有翠绿的叶子和亮粉红

太平洋岸边带雨点纹的石头，经受了千万年海浪的冲刷。

Rain speckled stones on the Pacific beach, washed by thousands of years of waves.

entrance, decked with pictures of men who have made great technologic advances in history. The lobby area is a tall ceilinged room where there are hundreds of computers, their screens flashing with blue, green and red figures, accompanied by tweeting signals and ringing alarms. I look at this beautiful building with its complex computers, and I exclaim to my friends around me, "oh this is just nothing, all it takes is millions of years of seawater sloshing over the shore, with intermittent bolts of lightning from the skies, and suddenly bingo, this all happens. It's just natural. It doesn't need any logic. It's all a matter of chance. There really is no goal in all of this. It just happened!"

We recognize a sea wave polished rock on a beach, a native made bowl, an artisan made jade carving, and a *well designed* computer office built by smart humans, because our brain recognizes each distinctive feature.

In the beauty and complexity that is in all of nature, we logically should recognize that there may be a super high logic and intelligence behind all of this. This beautiful complexity is present even in the so called "simplest" parts of our body (like, what is that?), any part of the body, or any cell of the body. We can't really say to ourselves, "Oh, it just happened, you see it's here, so it must have happened."

And where does all this logic and intelligence come from?

一个配备很多电脑的房间，要有多少次闪电才能创造出来呢？

A complex computer room: how many bolts of lightning does it take to create this?

2. 海滨漫步：寓言
2. Just Walking on the Beach: An Allegory

色的花朵。稍作清洗，冲刷掉树叶间的水草，我就可以把这件工艺品摆放在我家客厅里。我推断这个玉雕一定出自一位技术娴熟的中国工匠，甚至可能是很近期的作品。

突然，我惊讶地看见海滩的尽头有一座非常大的建筑物，我很好奇，想去一探究竟。我经过华丽的入口，看到一些人物画像，画中的人都曾经在历史上大大推动了科技的进步。大厅的屋顶很高，里面有数百台电脑，显示屏上闪烁着蓝色、绿色和红色的数据，还能听到啾啾的信号声和响铃提示音。我看着这座配备复杂电脑的美丽建筑物，向我身边的朋友们惊叹道："哦，这也算不得什么。只要海水冲刷岸边几百万年，天空间中有闪电击下，这一切就会忽然之间出现。这只是自然而然的。不需要任何逻辑。这全部是偶然发生的事。这一切其实没有什么目的。就是这么发生了！"

我们之所以能够辨认出岸边经过海浪冲刷的石头、土人造的碗、巧匠制作的玉雕，和聪明的人类**精心设计**的电脑办公大楼，是因为我们的大脑能够辨认出它们不同的特征。

自然界的一切都是美丽而复杂的，我们从逻辑上应该晓得其背后可能存在超高的逻辑和智慧。这种迷人的复杂甚至存在于我们身体中所谓"最简单"的部分（其实没有什么真是简单的），在身体的任何部分或任何细胞中。我们真的不能对自己说："哦，那是碰巧的，你看它就在这里，所以一定是这样。"

这些逻辑和智慧都是从何而来的呢？

翻译：Daisy Wang

3. Great Signatures: John Hancock and I. M. Pei

I. M. Pei is very famous. But who is John Hancock? Or maybe you know Mr. Hancock, and ask, "who is I. M. Pei?"

Americans instinctively know when someone *asks them for* a "John Hancock," but most non-Americans might be perplexed. John Hancock was a signer of the Constitution of the United States. He was not a great man by the standards of the original framers of the Constitution, but his claim on history is that he left a very large *artistic* signature, the *most obvious* signature of all the signers. From that point onwards, whenever people want you to sign your name in America, they will ask you to give your "John Hancock."

Creative work is, by definition, the work of a creator, an artist. And the creator often leaves some kind of signature on his work. For example, when we look at a painting that has many blues and greens of nature, we instinctively think of someone like Monet, because that is his "signature," practically instantly obvious.

I. M. Pei is one of the most famous creative architects of all time. I happen to have a marriage connection with him, so I have been particularly watching his career and creative work for many years. Mr. Pei grew up in Shanghai and lived for a significant time in New York, but his hometown was Suzhou, a very beautiful town in China, near Shanghai. Some of his most famous architectural buildings include the

3. 汉考克和贝聿铭的精彩签名

贝聿铭家喻户晓,但是约翰·汉考克(John Hancock)是何许人也?或许你知道汉考克先生,却不认识贝聿铭。

若有人向美国人**索取**"John Hancock",他们马上就知道是什么意思,而大多数非美国人就会感到困惑。约翰·汉考克是美国宪法的联署者之一。在多位最初制宪者当中,他不算是伟大的人物,但是他留下了大大的**艺术**签名,在所有签名当中最为**显眼**,让他青史留名。从那时起,在美国,别人想要你的签名,就会请求你提供你的"John Hancock"。

创意工作顾名思义就是创作者、艺术家的工作。创作者经常会在作品中留下某种形式的签名。举例来说,当我们看到一幅有许多蓝色和绿色的风景画时,就会立刻想到莫奈,因为这是他的"签名",非常明显。

贝聿铭(I. M. Pei)是有史以来最著名的创意建筑师之一。我碰巧和他有姻亲关系,因此多年来得以细致地观察他的事业发展和创意工作。贝先生在上海长大,后来长时间住在纽约,不过他的祖籍乃是与上海毗邻、美丽的中国城市苏州。他所设计的知名建筑物包括巴黎的卢浮宫和麻省的肯尼迪图书馆。传

Louvre in Paris and the Kennedy Library in Massachusetts. But there is a storied tradition that I. M. Pei loved to leave his signature *directly in* his architectural buildings. However, the signature is not usually immediately obvious, and you have to look for it.

One day I was strolling around Hong Kong Island, and examining the Bank of China building, one of his most famous buildings, and a critical architectural masterpiece of the Hong Kong skyline. To my great surprise, I "realized" that the building was actually an upside down "P", the "candles" on the side were basically the "i"s, and above these on the bold sides of the building were many "M"s; indeed all of the initials of "IMP". Not many people actually know that, at least not my many friends whom I asked. Did you?

In 2014 and 2015, I visited the Suzhou Museum which was designed by Mr. Pei, in his late years, as a tribute to his favorite city of Suzhou, his hometown. As I was wandering around, I decided that I *must* find his signature. It should be there in some way, according to the tradition. No one I asked knew what I was talking about, certainly not any local people.

After walking around for quite a while, I suddenly realized that the signature for his creative work was, as they say, "*hiding in plain sight.*" The inscription for the entire building was basically his name "Bei", but replicated again and again in various guises, written in a very artistic way (see the photos). The insignia of the museum itself, and

你认出了矗立在世界上最繁华的城市之一的这座经典建筑物上的签名吗？

Recognize the signature on this architectural masterpiece in one of the most exciting cities of the world?

3. 汉考克和贝聿铭的精彩签名
3. Great Signatures: John Hancock and I. M. Pei

闻贝聿铭喜欢在他的建筑物上**直接留名**,但是他的留名并不是一目了然,你须要仔细寻找。

有一天我漫步香港岛,审视他非常知名的、构成香港天际线的杰出建筑艺术品——中银大厦,我惊奇地发现这座建筑物实际上是个倒立的"P"字,边上的"蜡烛"基本上是许多个"I",而且,建筑物醒目的立面上有多个"M",正是他名字的首字母"IMP",代表贝聿铭。不是很多人知晓其中的秘密,至少我请教过的朋友大多数都不知道。你知道吗?

2014和2015年,我曾两次参观贝先生晚年设计的苏州博物馆,这是他奉献给他的故乡、他钟情的城市苏州的礼物。当我信步其间的时候,我**决心**找出他的"签名"来。按照惯例,它一定以某种形式存在那里。当我请教其他人时,竟无人明白我的意思,包括当地人。

徜徉了好一会儿后,我突然意识到这个作品的签名竟然"**躲藏在众目睽睽之下**"。整个建筑群基本上刻满了他的姓氏

苏州最美丽的博物馆,由当地最著名的子孙题名。
Suzhou's most beautiful museum signed off by its most famous native son.

all the buildings were built on this same motif: essentially the entire museum was "bei," "bei" and "bei," everywhere you look. But you have to look to see it. It is sort of like the phrase, "you see, and do not see" or "those who have eyes to see, can see." Or simply, we have to open our eyes to see many things, to be aware, and to be *willing* to see.

Our Creator is like that, He *wrote* all over creation, and it is "obvious" to those who have eyes to see; but "not obvious" to those who do not look, or do not want to look. He is there to be found if we want to find Him, His *signature* is all over the universe: His "John Hancock" and His "I. M. Pei" are everywhere.

A famous declaration by a great rabbi was:

"For since the creation of the world, God's invisible qualities - his eternal power and divine nature - have been clearly seen, being understood from what has been made, so that people are without excuse."

贝、贝和贝。
Bei, Bei and Bei.

3. 汉考克和贝聿铭的精彩签名
3. Great Signatures: John Hancock and I. M. Pei

"贝",不过伪装成多种不同的样貌,以非常艺术化的方式写出来(看图)。博物馆的标志和整个建筑群都遵循这个主旨,放眼望去,整个博物馆到处都是"贝"、"贝",还有"贝"。但是你还得仔细去查找,正所谓"视而不见"或者"有眼可见的人就能看见"。简言之,我们必须睁开双眼去观察事物,要有意识、很乐意地查看。

我们的造物主也是这样,他在所有的创造物上**写上**记号,用眼去仔细查看的人可以"一清二楚";不去看的或者不想看的人却如同"雾里看花"。你找他,他就会被发现,他的**签名**遍寰宇,他的"John Hancock"和他的"贝聿铭"无处不在。

一位伟大的犹太学者说:

"自从造天地以来,上主的永能和神性是明明可知的,虽是眼不能见,但借着所造之物就可以晓得,叫人无可推诿。"

翻译:Sonic

4. Shanghai Cell

I have been a scientist for more than 30 years. It never ceases to amaze me that perfectly logical and brilliant scientists would say quite matter-of-factly that evolution "caused" this or that to happen in history, as if it is a proven fact, well understood and generally accepted by everyone.

I often remark: I don't need you to explain how an ape evolved into a complex human being. Just explain to me how the first smallest cell ever came to being by "chance!" Just one little cell. The one little cell, the most "primitive" (to the uninitiated) cell, is actually as complex as the city of Shanghai!

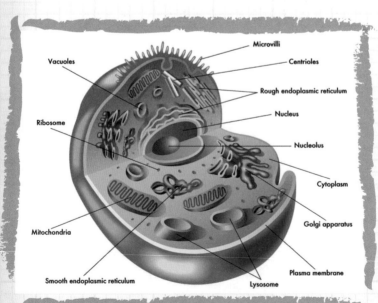

细胞解剖图
Anatomy of a cell

4. 上海细胞

我做科学家三十多年了,有件事一直让我好生惊讶。逻辑完美和绝顶聪明的科学家们总是煞有其事地宣称,是进化"促成"历史上发生这事或那事,说得好像这是个已被证明的事实,谁都明白,谁都接受了。

我经常说:猿类怎样进化成复杂的人类,我不需要你来解释。你就给我讲讲,第一个最小的细胞究竟怎样"偶然"诞生出来!就一个小小细胞。这个小细胞——在门外汉眼里最为"原始"的细胞——实际上复杂得像个大上海!

我们越了解细胞,就越知道它错综复杂得难以描述。所有的科学发现都揭示,细胞比我们先前所想的复杂得多。新的发现从来没有使细胞显得更简单。在我们看来,细胞总是越来越细致,越错综复杂,越有"智慧"。

我经常开玩笑说,要不是细胞这么复杂,就没有这么多好机会,让世界各地的学者们来到美国和其他科技先进的国家,拿博士学位,作博士后,甚至后博士后研究。好像设计细胞的这位"智慧"有意要让千千万万的学者能够以探索"惊人的细胞"为事业。

As we learn more about the cell, we learn that it is indescribably intricate. Every discovery of science reveals it is even more complex than we thought before. Every new discovery never makes it simpler. It is always more intricate, more complex and more "intelligent."

My usual joke is: were it not for the complexity of the cell, there would not be so many wonderful opportunities for scholars worldwide to come to the US, and many other technologically advanced countries, to obtain their PhDs, Post-Docs, and Post-Post-Docs. It is as if the Intelligence that designed the cell, designed it to allow literally tens of thousands of scholars to have a career exploring "the amazing cell."

And truly "the amazing cell" is like Shanghai: the simplest cell has the equivalent of phone lines, wireless connections, computer chips, cell phones, and emails zipping around inside. There are functions within the cell that serve as policemen, firemen, anti-pollution experts, garbage disposal staff, postmen and gardeners. In the cell there are features like street signs, traffic lights, superhighways, typhoon and tornado warnings, and fire alarms. And it all works beautifully: no garbage pileups, no traffic congestion, no plumbing backups, no water shortages, and no pollution in the cell. And it never stops working. And can we really imagine 50 trillion (50×10^{12}) cells in the body? Or 50 trillion Shanghais in each of us, all working in an amazing interplay and interchange.

Tell me how it all "came by chance?" One phone line at a time? One water pipe at a time? Or one traffic light at a time? Or did the Master Designer, of infinite wisdom and intelligence, put it ALL together as a PERFECT example of what can go right when all the pieces work together, and all the interlocking items fit perfectly? I prefer the Intelligent Designer over Illogical Chance.

4. 上海细胞
4. Shanghai Cell

  这个"惊人的细胞"的确就像大上海：再简单不过的细胞内，拥有相当于全上海的电话线路、无线网路、计算机芯片、移动电话和电邮，全部都同时急速地运转。细胞的种种机能各司其职，有警察、消防员、防污染专家、清洁工、邮递员和园丁。细胞中还有一些设施，如路标、交通灯、高速公路、暴风警告和火警系统。这一切都完美地运行，没有垃圾堆积，不会交通拥堵，没有管道阻塞，不会用水短缺，没有环境污染，而且永不停止工作。我们真的能想像身体中有五十兆（50×10^{12}）个细胞吗？或者说，我们每个人身上都有五十兆个上海，彼此之间还挺惊人地相互影响，相互交通呢。

  告诉我，所有这些都是怎样"偶然出现"的呢？一次一条电话线？一次一根水管？或者一时蹦出一个交通灯吗？还是有一位智慧无穷的设计大师，将这**一切**整合在一块儿，分工合作，各部分配合得天衣无缝，成为一**完美**的范例？我宁愿归功于满有智慧的设计师，甚于不合逻辑的偶然。

翻译：Joyce Wu

上海天际线
Shanghai Skyline

5. The Best Pain Doctor in the World

What if you woke up from deep sleep one morning, and found that the tips of your fingers were gone! Dr. Paul Brand, in his biography *10 Fingers for God*, described this scene when he checked up on some children who had leprosy. After a lot of investigation, he found blood spots leading from the bedside to a hole in the wall. Rats had come in the middle of the night, and nibbled on these tender fingers. The children's leprosy had damaged their nerves, and therefore they did *not* have pain, even when the rats were biting and nibbling away.

Dr. Brand (with Philip Yancey) wrote the book *Pain, the Gift that Nobody wants*. He described pain as a "gift from God," a concept that might be shocking to many people. But imagine yourself with leprosy, and your nerves are damaged by leprosy, so that you *cannot feel pain*. Or you have diabetes, and the nerves of your legs are damaged to not sense pain. Or you have been lying on your back for many weeks without changing your position, so that your back is numb, and you do not feel pain. In all these situations, because you have no sense of pain, you could get seriously injured, have infections, and even have parts of your body sloughed off without your even realizing it. Pain is indeed a gift to make us wake up, to escape from harm, or to see a doctor.

Three times in my life I developed sciatica, the compression of nerves coming from the spine. This was during a time when I had to travel 7 to

5. 世上最好的疼痛医生

某个清晨当你从酣睡中醒来,发现你的指尖不翼而飞!在保罗·布兰德医生的传记《十指为神》中,记述了他检查麻疯病孩子的情况。经过仔细的搜索,他发现斑斑血迹沿着床边直到墙角鼠洞。老鼠在午夜时分出没,啃噬了那些稚嫩的手指。麻疯病损害了这些小孩的神经,即使老鼠又啃又咬,他们根本不知道疼痛。

布兰德医生(和杨腓力先生)的著作《疼痛:健康失调的警讯》,英文原著的副题是"无人想要的礼物"。他把疼痛视为"上主的礼物",这个观念似乎很惊悚。但是试想想,如果

和保罗、玛格丽特·布兰德两位医生在西雅图。
With Doctors Paul and Margaret Brand in Seattle.

8 times a year on international trips, especially to mobilize doctors and nurses to go to China for the medical mission I had helped begin. My neurologist told me that the cause of the sciatica was most likely because I was sitting in a very cramped economy section of the airplane, *bent over* a laptop trying to clean up many careful notes from numerous meetings. He reminded me that this was the *worst position* to be in, and advised me in later life to bring a pillow for my back whenever I traveled. So this is why you keep seeing me with a *red bag*, which is really my support for my back, a pillow camouflaged as a bag. You may even have thought that my laptop was in this funny red bag: it is definitely *not* a laptop bag.

最有名的疼痛书。
The most famous pain book.

During these attacks of sciatica, I had literally *daily advice* from my kind eager medical mission staff, to see an acupuncturist, surgeon, chiropractor, physical therapist, pain specialist, etc or to use steroids, spinal injections of analgesics, etc. etc. I politely refused all the well intended advice, and settled on *total bed rest*, with the excruciating pain being my guide as to what *painful position to avoid* when I was lying down. I took minimal pain medication, but I maximally used pain as an indicator to give me relief and rest. During each of the 3 attacks, after one week or so, the pain gradually subsided, to everyone's surprise. Basically after these three attacks, and especially later with the help of my little red pillow bag, I have been free of attacks for more than a decade. Pain was clearly a gift for me in this case. And I narrowly escaped the surgeon's fine

5. 世上最好的疼痛医生
5. The Best Pain Doctor in the World

你患有麻疯,你的神经被麻疯损坏,以致你**感觉不到疼痛**。或者你患有糖尿病,你腿部神经被破坏,对疼痛毫无知觉。或者你卧床数周,不改变姿势,你的背部就会麻木,丧失痛感。凡此种种,因为你感受不到疼痛,就可能会受到严重伤害、感染,甚至身体某部分在不知不觉中腐烂掉。疼痛实在是个礼物,让我们警觉,让我们规避伤害,或者求医。

我曾经三次因脊椎神经受压迫以致坐骨神经痛,都发生在我不得不每年远遊国外七至八次的时候,尤其是在动员医生、护士去中国参加我所帮助启动的医疗服务期间。神经科专家告诉我,我的坐骨神经痛主要是因为我坐在狭窄的经济舱,**弯腰**用笔记本电脑整理众多会议的注意要点。他提醒我这是**最不好的姿势**,建议我在以后的旅行中带个枕头作为靠背。所以你会经常看到我带着一个**红色的包**,它是我背部的支撑,是伪装成包的枕头。也许你以为我的笔记本电脑就装在这个有趣的红包里,但是它其实**不是**电脑包。

坐骨神经痛发作时,我**每天**都得到我的热心医疗服务团队伙伴的**建议**,要我去看针灸师、外科医生、按摩师、物理治疗师、痛症专家等,或者应用类固醇、椎管注射镇痛药,诸如此类。我婉拒了所有的善意,决定**彻底卧床休息**。在我躺下的时候,剧痛提示我**避免引起疼痛的姿势**。我服用最小剂量的止痛药,反而尽量利用疼痛作为提示我放松和休息的讯号。这三次坐骨神经痛的每一次发作期间,让人惊讶的是差不多一周左右,

hands and scalpels. Dr Brand was right, nobody wants pain, but it is a gift we can use.

In fact there are very rare children born with *no pain sensation*, and they commonly will injure themselves extensively. It was Dr. Brand's dream to find a way to actually *create pain sensation*, in order to provide the *protection* necessary. That is really difficult, since we automatically assume our sensory nerves know what they are doing, and if there is a problem, it is commonly the opposite situation, having too much pain, so we are trying to reduce the sensation, rather than increase it. It is still not possible to create pain sensation. Not many researchers probably think that creating pain is a *career path*, again a gift that nobody wants. And this genetic disease reminds us that having no pain is actually a curse also, not a blessing.

Paul Brand was the most gentle doctor that I have ever met. Sometimes we joke that surgeons can be very aggressive, and even abrasive, because they have often to be quick in mind and action. I know this from personal experience, since I abruptly stopped my budding surgery career when I came under the direction of just such a person in my internship year. But here was Dr. Brand, totally against stereotype, carefully attending the *wounds of many leprosy patients*, cleaning their wounds, or binding their damaged feet. Or advising me about medical missions, in his very British, fatherly and kind manner.

When I started my medical mission during my first early retirement, I was anxious to find a medical person who could be my trusted advisor. I had already read books written by Dr. Paul Brand, and from his books I could sense that he was a very sensitive, logical and godly person. His thoughtful book *Fearfully and Wonderfully Made* was a great reminder of the beauty of God's creation of the human being, and how complex

5. 世上最好的疼痛医生
5. The Best Pain Doctor in the World

疼痛就逐渐缓解。基本上在这三次发作以后，尤其是后来在我的小红枕包的帮助下，坐骨神经痛已经十多年没有发作了。在这种情形下，疼痛毫无疑问是我的礼物。我惊险地躲过了外科医生精巧的手和解剖刀。布兰德医生是对的，没有人想痛，但它是我们有用的礼物。

事实上有极少的小孩是生而**无痛觉**的，他们往往会时不时地伤到自己。布兰德医生的梦想是找到**创造痛觉**的方法，目的是为人体提供必要的**保护**。这太难了，因为我们理所当然地认为感觉神经知道自己该司何职，而且我们遇到的问题通常都是恰恰相反的，就是疼痛难忍，因此我们尽力减少这种感觉，而不是增强它。迄今为止人类仍然未能创造出痛觉。没有研究员会以制造疼痛为**事业**，他们还是认为疼痛是没有人想要的礼物。这种遗传病也提醒我们，没有疼痛其实是诅咒，而不是祝福。

保罗·布兰德是我遇到的最温和亲切的医生。我们有时开玩笑说外科医生可以非常进取，甚至粗鲁，因为他们经常敏于思考，捷于行动。这观点来自于我的亲身经历，我实习期间刚好遇上一位这样的指导老师，因此我仓促终止了我刚刚开始的外科职业生涯。但是布兰德医生彻底颠覆这种刻板印象。他悉心护理**众多麻疯病人的伤口**，清洗患处，包扎烂脚。又以他英伦式、父亲式、和蔼可亲的态度给我医疗服务方面的建议。

在我第一次提早退休，刚开始医疗服务的时候，我迫切地想找一位医学界人士作为我可信赖的顾问。我读过保罗·布兰

and wonderful the entire design reflected God's genius. Further, he had discovered, by careful observation and experimentation, that the disfigurement of leprosy was not the direct result of the leprosy disease, but a side effect of the nerve damage and poor pain sensation. This was a revolutionary finding that dramatically altered the management of leprosy, and I was really impressed.

I wrote Dr Brand *out of the blue*, since we did not know each other, and I had nobody even to introduce me to him. So I just took the direct, likely brash route, and simply wrote him. But he readily accepted my request to have him be my mentor, and every year I would fly on a "pilgrimage" to his home, so conveniently Seattle, my mother's birthplace and our own future home. And every time we met it was just a wonderfully meaningful experience, to meet him, and Mrs. Margaret Brand, also a doctor, who specialized in treating the eyes of leprosy patients.

Paul was a man of few words, and when I had a question about my mission, he would listen very intently, raise his eyebrows, frown, or have a faint smile of approval. Mainly from these *faint gestures* and body language, I learned a lot! As I reported what I was doing, he would *gently steer* me into the right directions, without once having to raise his voice or directly argue with me. Margaret however was much more vocal and was a great counterpoint during this mentoring period. Mentors truly come in all shades, and they are all extremely helpful in different ways! At this point in the story, you might be thinking it strange that my mentors were leprosy specialists, but there is likely a deeper reason.

Ever since I was a child, I have had a fascination with leprosy. My father had many missionary friends, including a Dr. Fraser who was a leprosy specialist. We used to go and visit him on an island especially

5. 世上最好的疼痛医生
5. The Best Pain Doctor in the World

德医生写的书,从他的书中我感受到他敏锐、有逻辑,而且是信奉上主的人。他的著作《我受造奇妙可畏》寓意深刻,令人回想上主创造人类的美好,以及整个设计的复杂和精巧体现了上主的大能。后来他通过仔细观察和实验,发现麻疯病人的身体损伤不是由麻疯病直接导致,而是神经损坏和痛觉缺失的不良后果。这是一个革命性的发现,大大改变了麻疯病的治疗方式,也给我很深刻的印象。

我们彼此并不认识,也乏人引荐,所以我**冒昧**地写信给他。虽然我行事唐突,但是他欣然接受了我的请求,作我的导师。每年我都飞到他家"朝圣"。他的家刚好在西雅图,这里是我母亲的出生地,也是我们未来的家。玛格丽特·布兰德太太也是一位医生,专长于麻疯病人的眼科治疗。每次去见他们都是令人愉快、意义深远的经历。

保罗是个寡言的人。当我提到关于医疗服务的疑问时,他倾听、扬眉、皱眉,或者轻轻微笑认可。仅从这些**细微的姿势**和肢体语言,我就领悟到许多!当我汇报我的工作时,他会**轻轻地校正我的方向**,从不提高语调或者和我直接争论。在请益过程中,玛格丽特经常发言,形成了对比。导师的风格千变万化,全都能令人获益良多!故事讲到这里,也许你会感到奇怪,我的导师居然是一对麻疯病专家,不过这是有更深入的原因的。

我还是小孩的时候,对于麻疯病非常着迷。我父亲有许多传教士朋友,其中弗雷泽医生就是一位麻疯病专家。我们经常

built for leprosy patients. The beaches there were pristine, since *no one* really went to visit the island. We had a great time swimming on the beaches, and also visiting with the leprosy patients. I learned not to be scared of the disease nor the patients. These experiences probably helped me in my future calling to medical missions, and helped me focus on leprosy as a fascinating disease.

Leprosy is indeed the prototypical disease that reflects the love of Christ. He reached out to them, He *physically touched them* at a time when no-one dared to be even close to them, and He healed them. Among the most despised group of people in the world, leprosy patients have, for thousands of years, encountered discrimination and marginalization. But Christ's model has inspired thousands of medical missionaries to find the sickest, poorest and most neglected of the world, to help them and encourage them. Christ was really the supreme *pain doctor*. And it is with great joy and pride that I was able to associate for many years with the best *human* pain doctor ever.

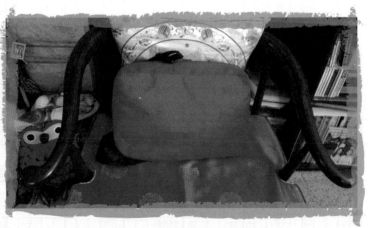

心爱的红包，多年的伙伴，疼痛礼物的提醒。
Favorite red bag, my buddy for years, as a great reminder of the gift of pain.

5. 世上最好的疼痛医生
5. The Best Pain Doctor in the World

去专门为麻疯病人开辟的岛上看望他。因为根本**没有人**愿意踏足,所以岛上海滩的原生态得以保留。我们很享受在海滩游泳,也去看望麻疯病人。我认识到不必害怕麻疯病和麻疯病人。这些经历可能有助于我后来蒙召投身医疗服务,也帮助我关注麻疯病,视之为一个引人入胜的疾病。

麻疯病是一个最能体现基督之爱的疾病。基督向麻疯病人伸出双臂,**亲手触摸他们**。在那个时代甚至没有人敢靠近他们,而基督却治愈他们。麻疯病人是世上最被蔑视的群体之一,千百年来遭到歧视和边缘化。但是基督的榜样启发了千千万万的医疗传教士,去寻找病情最重、最贫苦和最被世人忽视的麻疯病人,帮助他们,鼓励他们。基督才是至高的**疼痛医生**。我能和最好的**人类**疼痛医生保持多年联系,让我十分高兴,十分自豪。

翻译:Sonic

6. Congratulations, Your Baby Has a Mutation?

I have been a neonatologist-pediatrician for over 30 years, and I have seen hundreds of mutation affected infants, and read about thousands more in the medical literature. And I have never seen a real live "good" mutation. (Note: a neonatologist is a doctor who specializes in neonates, or newborn care).

In fact, the medical literature is full of reports showing how even a small one amino acid mistake / mutation in the phenomenally complex DNA can cause a profound problem. So how is it the biologist, who has never seen a single human patient with mutation, can claim, or "fantasize", that mutation is the essence of evolution, which propels organisms to increasing complexity and "advance" to a "higher" species.

In fact, if evolution (or "macro"-evolution to be precise) really works in the real world we would need to have repeated "good" mutations, that, repeatedly "advance" the species so that, bingo, one day, a "super species" now appears, and we have jumped from one species to another, such as an ape supposedly to a man.

Imagine a neonatologist or pediatrician smiling broadly one evening, and strolling into the delivery room of an anxious mother who has just delivered a baby and awaiting the result of the doctor's physical examination. The good doctor announces with a wide grin, "Mrs. Jones, congratulations. You have delivered a baby with a great mutation!"

6. 恭喜您，
您的小宝宝有生理突变？

身为新生儿专科医生，三十多年来，我曾见过数以百计的婴儿受到生理突变的影响，并在医学刊物上读到数以千计这样的报告。可是我从未见过一宗活生生的「良性」突变的真实例子。

其实，医学刊物上反而是充满了这样的报告：在异常复杂的 DNA 中，甚至一个小小氨基酸的错乱或突变都会带来极严重的问题。所以，生物学家如果连一个有生理突变现象的病人也没见过，又怎能宣称或"幻想"突变是进化的精髓，认为突变足以推动生物更趋复杂，"进步"成"更高等"的物种？

其实，如果进化（更确切的说法是"宏观演化"）在现实世界中真有其事，那么我们就应该去重现那些"良性"突变，令物种不断地"进步"，以至有一天，你瞧，出现了一个"超级物种"，我们就这样从一物种跃变成另一物种，正如从猿类变成人类一样。

想像一下，某天晚上一位新生儿专科医生或儿科医生面带愉悦的笑容，走进产房，见到那位刚才分娩的产妇，她正等着

Sorry, sci-fi folks, this NEVER happens. The word "mutation" in real life is a terror loaded word. We pediatricians say the word softly, quietly, and carefully, and try very hard to use other words around it to "soften" the deadly blow that the mother will sense immediately. No, Alice, there are no good mutations in real life. Maybe in Wonderland.

In an infinitely (and still barely understood) complex strand of DNA, which truly is a mind-boggling work of science and art, everything seems to be there "for a purpose" (we call such "purpose driven" biology, "teleology" [not theology]). And it is always amusing to watch confident scientists belittle this or that new finding in biology when it is first identified, as something "useless", but having to eat their words later. Witness the odd use of the word "vestigial" or "junk" sprinkled liberally as adjective for substances of unknown function at first, only to be reported later as "amazing", or "critical" components of an "intricate system" of controls and programming.

镰状红血球。
The Sickle Cells as sickle forms.

6. 恭喜您,您的小宝宝有生理突变?
6. Congratulations, Your Baby Has a Mutation?

医生给婴儿作身体检查的报告。这位好医生露齿而笑,宣布说:"钟太太,恭喜您!您的小宝宝有很好的生理突变!"

抱歉,爱看科幻小说的朋友,这种事**绝不会**发生。在现实生活中,"突变"是个带着恐怖色彩的字眼。每当提到这个字眼,我们儿科医生总是尽量轻柔谨慎,而且费尽心思地用其他一些"软性"字眼来包裹它,希望那位母亲不至于太承受不了那致命的打击。不,爱丽丝,现实生活中并不存在良性的突变,虽然在你梦遊的仙境里或许有。

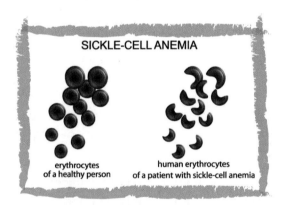

DNA 中一个微小的改变,一个突变,使红血球变成镰状,成为严重的致命疾病。
One small change in DNA, a mutation, causes the red cell to become sickled, a serious potentially life-threatening disease.

错综复杂(科学家至今仍对之所知甚少)的 DNA,实在是令人惊叹、难以想像的一种科学与艺术之作,其中任何一部分的存在似乎都是"有目的"的(我们称之为"目的导向的"生物学,或"目的论")。有趣的是,每当在生物学上有新的发现,一些自视甚高的科学家往往喜欢任意贬低,说这说那是"无用的"东西,但是到后来又不得不收回自己先前所说的话。人们喜欢随便用"残留的"或"垃圾"来形容那些最初未知其功能的东西,后

With the amazingly complex molecular structure in every biologic cell, with everything in it seemingly there for a purpose, it is really little wonder that any "mistake" or "mutation" will result logically in a turn not for the better, but for the worse, which is what happens in real life! "Wait", you say, "I know of one good mutation, at least. It's the sickle cell trait that provides protection against malaria." Let's grant that there might be some "advantage" in that regard (and the data for that are not strong, usually simply a rehashing of old poorly controlled semi-epidemiologic studies). However, that needs to be balanced against the fact that sickle cell disease is often lethal! How "beneficial" is a condition that could result in early death. Imagine again, the pediatrician smiling broadly, and announcing to the mother, "congratulations, you have a sickle cell mutant baby!"

Furthermore, millions of US dollars are spent by the National Institute of Health, and the Sickle Cell Foundation to, get this, find ways to genetically manipulate the sickle cell mutation, or thwart its detrimental effects. If it is so wonderful, why waste all that effort to ameliorate the mutation, Alice?

Furthermore, how does this argument help in the (macro)-evolution theory? That is, how does this help lead to a super species, Alice? Imagine a sickle cell ape, which leads to a (sickle? or non-sickle?) mutant human? Alice, are you still there?

Mutations, Alice, are not the mechanism for evolution, but its death knell.

所有婴儿都应该生来没有突变，这样每个人都会非常开心。
All babies should be born with no mutations: everyone will be very happy. (Photo from JPW, California)

6. 恭喜您，您的小宝宝有生理突变？
6. Congratulations, Your Baby Has a Mutation?

来却又报告说，这其实是某个"精密"的监控系统或流程中"奇妙"或"关键"的零件。

每个生物细胞中的分子结构都异常复杂，其中每一部分的存在似乎都是为着某种目的，所以若有任何"错乱"或"突变"，自然都不是带来改善，而是带来恶化的转变——这正是现实生活中所见到的情况！你或许会说："且慢，但我至少知道有一种良性的突变，就是镰状红血球特征，有助于防止疟疾。"姑且假定在这方面有一些"益处"（有关的资料并不够坚实，通常只是重新包装过去一些有关半流行病学的不精确研究，换汤不换药），然而同时也要知道，镰状红血球病症往往是致命的！一种导致早夭的生理情况又怎能说是"有益的"呢？请您再次想像，那位面带笑容的儿科医生对刚分娩的母亲宣布说："恭喜您！您的小宝宝有镰状红血球突变！"

况且美国国家卫生研究院和镰状红血球基金会正花费数以百万计的美元，就为找出能从遗传学上控制镰状红血球突变的方法，或阻挠其有害的影响。爱丽丝，如果这种突变真是那么好，又何必枉费那么多力气去改善？

再说，良性突变的说法又如何有助于（宏观）演化的理论？或者说，如何有助于超级物种的诞生？难道是一只有镰状红血球的猿，跃变成一个有（镰状？或非镰状？）突变的人？嗨，爱丽丝，醒醒吧！

爱丽丝，突变并非进化的机理，而是进化的丧钟。

翻译：Joyce Wu

7. You Might Get a Good Mutation!?

As a neonatologist, a pediatrician who specializes in the care of premature babies, it has been my privilege for decades to be overseeing the care of thousands of premature infants in several nurseries. In a normal clinical setting, commonly we bring in X-ray machines in order to determine if there are problems in the lungs, abdomen, and bone (my area of research). When the X-ray machine rolls in, and is positioning itself to start the X-ray process, usually all the doctors and nurses step back 6 feet or more away from the baby, since no one wants to be irradiated.

From the viewpoint of general biology, we have been taught for generations that mutations are the essence of Darwinian evolution. Over millions and millions of years, many mutations have occurred, and bingo, the organism becomes more advanced, and that is how evolution ("macro-evolution") is supposed to work, developing an ascending order of complexity. Presumably, many, many mutations occur, so that so-called "lower species" develop into "advanced species", through the varied mutations that "advance" the species.

And how do mutations come about? Mutations come about, from radiation or noxious chemical agents in the environment. These agents stimulate the DNA to mutate, and with each mutation comes the "potential" for "advancing" the species.

7. 或许你会得良性突变呢！

身为新生儿专科医生，专长于照顾早产儿，在过去几十年里，我有幸在不同的医院里照顾过数以千计的早产儿。一般情况下，我们会采用 X 光机来检查婴儿的肺、腹腔和骨骼（这是我专门研究的范畴）。当 X 光机被推进来并架设好，开始拍摄时，通常所有的医生和护士都会从婴儿身边退后 6 英呎或更远，因为没有人想被辐射波及。

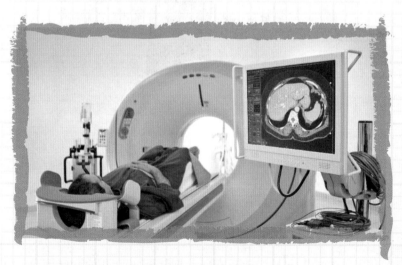

当病人被照射时，X 光机技师站在辐射屏障后面。
X-ray technicians stand behind radiation barriers when patients are radiated.

So, one day when the X-ray machine was being positioned to X-ray the infant in the nursery, I called out to my team of doctors and nurses, "don't leave! You might get a good mutation!" No one of course, believes this in real life. And there is a team chuckle, since everyone in real life medicine knows that mutations are not "good", and especially mutations from radiation or noxious agents. So how is it, that mutations in abstract theory have become the mechanism for advancement?

对病人（尤其是婴儿）照 X 光时，医生会尽量降低辐射量。
Doctors are careful not to give too much radiation especially to infants.

Damaging an Intricate system

The medical world is full of such examples of mutation that cause misery, especially in the pediatric wards. It is fascinating that in the real world of medicine, on a practical level, no one believes that mutations caused by radiation or noxious agents are good for you. But in the

7. 或许你会得良性突变呢！
7. You Might Get a Good Mutation!?

在普通生物学里，我们一代又一代被灌输：达尔文进化论的精髓就是突变。即是在过去的亿万年里，发生了一个又一个突变，碰巧生物得以进化。生物构造的复杂程度由此不断提高，进化（"宏观演化"）就是这个意思。据推测，突变不断发生，因此所谓"低等物种"就通过各种突变"进步"成"高等物种"。

突变是怎么发生的呢？它是通过环境里的辐射和化学毒物产生的。这些成分刺激 DNA 发生突变，而每一次突变在某些人看来都有让物种"进步"的"可能"。

但是，当 X 光机被推进婴儿室要为婴儿拍摄，如果我对团队中的医生和护士说："请留步！或许你会得良性突变呢！"当然没有人会相信。他们只会忍不住嗤笑，因为在现实生活中，所有医疗人员都知道从来没有"良性"突变，尤其是由辐射和毒物所引起的。如此说来，突变凭什么会在抽象的理论中变成生物进化的机理呢？

破坏一个精巧的系统

医学界充满了由突变酿成的悲剧，幼儿受害尤深。发人深思的是，在实践的层面，医学界没有人认为由辐射和毒物引起的突变对人有好处。然而，在那些坐在扶手椅上、对着电脑的生物学家的脑海里，"良性"突变却是有可能发生的，而且这个信念根深蒂固。（虽然他们已经在果蝇身上做过无数次放射实验，但还是不曾真正观察到"良性"突变导致进化。）他们

armchair world of the biologist sitting at a computer, who has never actually observed a "good" mutation that advances the species, (even from thousands of experiments on drosophila flies being bombarded by radiation), there is this very firm belief that it could happen. It might have happened in the past, it should have happened in the past, and it will happen again. Such a belief in an abstract theory is impressive, given the number of people who still believe in this decades old never proven idea.

Living in the real world of medicine for more than three decades has made me instead a firm believer that, the complex DNA in all biologic systems is so intricate and sophisticated, that any mutation in, or any damage to, this superbly designed system presents a tremendous danger. It is like throwing a stone at a beautifully complex glass structure, and hoping it would make it become prettier.

Increasingly, I stand in awe, as we begin to appreciate the ever complicated understanding of this complex system of organization. The Designer knew what He was doing. And we irradiate and add noxious chemicals to our detriment. Certainly not to our advancement!

7. 或许你会得良性突变呢！
7. You Might Get a Good Mutation!?

认为"过去可能发生过"、"过去应该发生过",从而推断"将来还会发生"。这么多人固守着几十年来都不能证实的抽象理论,实在让人震惊。

通过三十多年的实际临床医学研究,我实实在在地相信,生物系统中的DNA如此精密复杂,任何突变或破坏对于设计如此完美的系统都只能产生极大的危险。就像用石头去砸一个美丽复杂的玻璃制品,难道会令它变得更加美轮美奂吗?

当我对生物系统之复杂了解越深,我对它的设计就越发惊叹。这个系统的设计师其实是胸有成竹的。辐射和化学毒物只能损害我们,绝不肯使我们进化!

翻译:权克明

我们若用石头砸向美丽的玻璃制品,结果会怎样呢?
What would happen if we threw a stone at a beautiful glass structure?

8. Darwin's Eye: Absurd in the Highest Degree

The little old lady in a small village in southwest China was amazed. She cried out, "I can see, I can see!" Even though she couldn't see things very clearly, she now recognized her family whom she hadn't seen for years. The eye team from our medical mission had just performed a minor miracle again. Just like similar minor miracles again and again, over nearly 2 decades of work, in little villages where high tech medical care was difficult to reach. And it was always such a great joy to see the responses, the smiles, and the appreciation which the families had for the team and it's efforts. In point of fact, in today's modern high tech hospitals, the most common critical eye problem these villagers had, cataracts of older age, is easily treated. But not if you are old and living in poverty in a remote village, in many mountain areas of the world.

For the medical mission team, it was easy to associate the joy of this woman, with the joy of the blind man healed by Christ, who was vividly recorded for us thousands of years ago. The critics of Christ at the time, who did not appreciate his miracles, were harassing the formerly blind man, after the healing. They wanted to see if there was some fault in his account, and especially if there was something they could accuse the man who had healed him. But, the ex-blind man exclaimed, "I don't know much about this man (Christ), but one thing I know, whereas I was blind, now I see!" Truly, he wasn't really sure how it could have happened, nor

8. 达尔文的眼睛：荒谬至极

中国西南部小乡村的一位老妇人简直惊呆了，她大叫："我看到了，我看到了！"尽管她看得还不是很清楚，但是现在她可以认出她多年未见过的家人。我们医疗服务的眼科团队刚刚又行了一个小奇迹。在这些高科技医疗服务难以触及的小乡村，在我们近二十年的服务里，这样的小奇迹一次又一次出现。看到村民的反应和微笑，以及那些家庭对医疗队所付出的努力表达感激，令我满心喜悦。事实上，现今在技术雄厚的现代化医院里，这些村民当中最常见的严重眼疾——老年性白内障是容易治疗的。但是如果你生活在世界各地的偏远山区小村落，既老且穷，那可就难了。

医疗服务队很容易从老妇人的喜悦联想到被基督治愈的盲人之喜悦，这事迹在几千年前就已经为我们生动地记录下来了。那时候，基督的反对者不但不欣赏他所行的奇迹，反倒去挑剔被治愈的盲人。他们企图找出盲人的陈述中的破绽，特别想找出点什么证据好控告治愈他的那人。但是，被治愈的那盲人坚称："我对这人（基督）所知不多。有一件事我知道：从前我是眼瞎的，如今能看见了！"是的，他并不清楚奇迹是怎么发

每年很多宗视力手术
Many teams of eyesight operation per year

国际医疗服务机构 MSI 眼科组在贫困地区施行白内障手术，患者感受到戏剧性的变化。
Medical Services International, MSI Eye team performs cataract surgery in poverty areas, with dramatic results.

眼科项目负责人 George Chin 医生
Eye project leader - Dr. George Chin

could he discuss the philosophy or theology of what happened, but he knew for sure what had happened to his eyesight. Indeed, having eyesight is a world of difference from having no eyesight, and those of us who have normal vision, take our eyes for granted, and oftentimes forget how truly amazing these eyes really are.

We all know that science has advanced in leaps and bounds over the last century. When I was a teenager studying in high school, half a century ago, the science books used to call the cell, "the simple cell," because truly there was so little really known about it. Well, we know now that the cell is definitely *not* simple. In fact it is so complicated it "hurts your brain" to think of it. And you can guess, 150 years ago, when Charles Darwin was roaming around in the Galapagos Islands, his knowledge of

8. 达尔文的眼睛:荒谬至极
8. Darwin's Eye: Absurd in the Highest Degree

生的,也不会探讨这事的哲学或神学意义,但是他确知他自己的视力变化。事实上,和盲人相比,有视力简直是另外一番天地。我们有正常视力,对于我们的双眼习以为常,常常会忘记眼睛是多么的奇妙。

我们知道科技从上一个世纪以来飞速发展,日新月异。半个世纪前,我还是在高中读书的少年,科学书刊会称细胞作"简单的细胞",因为那时大家对细胞所知甚少。然而我们现在知道细胞可**不是**那么简单。事实上它是如此复杂,就是绞尽脑汁也想不明白。你可以想像,150年前,查尔斯·达尔文环遊加拉帕戈斯群岛的时候,他的科学知识(特别是生物学)是何等浅薄,连我们今日的常识都不具备,叫人难以想像。这差异的程度就像白内障让人失明,而摘除白内障后重见光明一样。

可以想想,那时没有电子显微镜,没有细胞功能的知识,没有基因和DNA的知识,没有基因在分子层面如何发挥作用的知识等等。那时候知识匮乏的程度对于今天一般的中学生而言是难以想像的。两个时代对于事物的理解有所不同,让我们明白达尔文时代的人为什么会想出我们现在认为不可能的**哲学理论**。身体系统之复杂在那时是完全未知的;现在看来,身体的**任何**部位、任何组织、任何细胞都是非常复杂的,远远超越了那个时候的人的想像。

我带着那样的观点读过达尔文的著作《物种起源》,我很理解他试图搞清楚生物学上究竟发生什么事。事实上,他和他

science, especially of biology, was so rudimentary, that we have difficulty even imagining that lack of today's ordinary knowledge. It is close to the difference between cataract affected blindness, and cataract removed eyesight.

Imagine, there was then no electron microscope, no knowledge of cellular functions, no knowledge of genes and DNA, no knowledge of how genes work at the molecular level etc. etc., something unimaginable to the average middle school student today. The difference in understanding gives us an appreciation of why people of Darwin's generation would come up with *philosophical theories* that are now known not to be possible. The complexity of the body systems was then totally unknown, and truly today *any* part of the body, any tissue, any cell is known to fantastically more complex than anyone could have even dreamed of at that time.

I have read Darwin's book *The Origin of Species* with that perspective in mind, and I can sympathize with his trying to understand what was actually going on in biology. Literally, millions of obstacles, filters and clouded vision were between him and the objects of his study. And so, I have always been very empathetic that his inquiring mind tried really hard, and am impressed that he could see, even dimly, that there were certain things that he intuitively knew had *huge problems*. When he looked at the eye, he realized that the complexity of function that was known even at that time, was truly astonishingly frightening to him. He wrote, in his book, "to suppose that the eye with all its *inimitable contrivances* for adjusting the *focus* to different distances, admitting different *amounts of light*, and correction of *spherical and chromatic aberration*, could have been formed by natural selection, seems I freely confess, absurd in the highest degree." I admire him for his astute observation, perfectly correctly, that natural selection, a key principle in his theory, was totally stumped. He

8. 达尔文的眼睛：荒谬至极
8. Darwin's Eye: Absurd in the Highest Degree

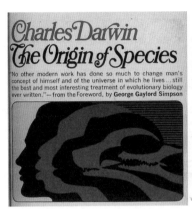
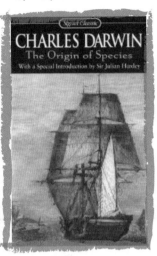

查尔斯·达尔文的大作，尽管他缺乏现代学术资格和理解力。
Charles Darwin's dramatic book, in spite of lack of modern scholastic credentials and understanding.

的研究对象之间，隔着无数的障碍、滤镜和模糊视线。他的探究精神使他努力不懈，可是他的直觉却依稀觉得这当中存在着**巨大的问题**，我对此感同身受。当他研究到眼睛，即使在那个时代，他就已经被眼睛功能之复杂深深震惊了。他在书中写道："眼睛有着**独特的设计**，用来调节**焦距**、控制**采光量**，及修正**球面像差和色差**。**如果认为这些都是物竞天择所致，我坦白承认，这个观点荒谬至极。**"我欣赏他机敏的观察，确实是这样的，他那套理论的核心——物竞天择难以自圆其说。他坦承那是"荒谬至极"的。

达尔文这么谦虚，就是今天也让人觉得难能可贵。他被眼睛之复杂震惊了；其实整个身体在系统、器官、组织、细胞各个层面，有**数不清**的例子都可以**完全适用**这个评语。他意识到

freely confessed that it was "absurd to the highest degree."

Darwin's humility is refreshing even today. As he was shocked by the complexity of the eye; to use a common jargon, there are *zillions* of other examples all over the body, of systems, of organs, of tissues, of cells where this comment would be *totally applicable*. He recognized that there were huge holes in his theory, "absurd to the highest degree," which holes are today fantastically *bigger and much more impossible* to be covered over by a statement that "natural selection" will "somehow" take care of it. Somehow is the key. In fact, "*how*" is the province of science, old and modern, and if we cannot demonstrate the "how" we have lost our job. Theories are nice, but over 150 years, the difficulties have gotten infinitely worse, and this theory totally fails, because we have no way of showing "how" it could have happened. "*The devil is in the details*," as they say. Indeed, "how" could the eye have formed? "How could it have evolved?" The silence is deafening.

My personal opinion is actually quite simple: Darwin had gone to seminary, and so he was schooled in philosophy. But he dropped out of seminary. Then he was sent to medical school, but he dropped out also. And he had no postgraduate training, no PhD. He was from a rich family, the cultural elite of the time, he did not have to earn a living, and basically, he was trying to figure out *philosophically* who we are on this earth. So, with family money and connections, he embarked on this journey, basically around the world, to test his thinking. Of course, he had no concept of modern physiology or molecular biology, no epidemiologic training, no statistics training. By today's standards I suppose he was at best like a modern 7th grade level student in science. But he had to make do with what he knew at the time, and his perception about his limits is revealing.

8. 达尔文的眼睛：荒谬至极
8. Darwin's Eye: Absurd in the Highest Degree

他的理论有很大的漏洞，"荒谬至极"。这些漏洞在今天看来可就**更大**了，在今天**更不能**靠"物竞天择不知怎么的发生了"这种论调来掩饰。"不知怎么的"是个关键。事实上**"怎样发生"**是科学界关注的重点，无论过去或者现在。如果我们不能展示"怎样发生"，恐怕已经被开除了。有理论本来是好事，但是150年过去了，雪上加霜，因为我们没办法展示物竞天择怎样发生，所以这个理论彻底破产了。谚语说："**魔鬼躲在细节里**"。眼睛"怎样"形成？它是如何进化的？答案可能是沉默，却震耳欲聋。

我的个人观点很简单：达尔文上过神学院，受过哲学教育，但是他从神学院退学了。接着他进了医学院，又退学了。他没有接受过研究院的培训，没有博士资格。他出生于富家，是那时的文化精英，不用谋生。基本上，他只是尝试**从哲学角度**探讨"在这地球上我们是谁"。于是，他靠着家人的经济资助和人脉关系，开始环游世界，去验证他的想法。当然，他没有现代生理学或分子生物学的概念，没有流行病学的培训，没有统计学的学习。以现在的标准来看，我认为他的科学知识充其量相当于初中新生的水平。但是他还是运用当时他所知道的去分析，也感受到了他自己知识上的限制。

若用现代科学去研究眼睛，我们现在拥有的知识也许只是一百年后的小部分，这让我们学会谦逊，懂得承认我们知识的有限。就是现在，我们也承认眼睛是复杂至极的系统，它是如此巧妙的构建，如此复杂的组合，在数学上如此精准，只有一

When we look at the eye today in modern science, which might only be a fraction of what we might know a hundred years *later*, it behooves us to be *humble*, and to appreciate the limits of our knowledge. Even now, we should recognize that the impossibly complex system that is the eye, is so wonderfully constructed, so complexly integrated, and so mathematically precise, that a *vastly superior intelligence* who designed it all, is still our best conclusion. And actually, it is the best conclusion for everything else in the body. A Creator, not a new idea, but an idea from the beginning of time, *in nearly all cultures and races* of mankind. A conclusion that ties together many disparate facts, and gives us a reason for our own existence.

8. 达尔文的眼睛：荒谬至极
8. Darwin's Eye: Absurd in the Highest Degree

位**超高的智慧**才能设计出来，这依然是我们的最佳结论。事实上，关于身体的任何部分也是如此。"造物主"并不是新的想法，而是从亘古就有的观念，并**存在于几乎所有不同文化和种族**的人心中。这个结论串联起许多不同的现象，也解释了我们自己存在的缘由。

翻译：Sonic

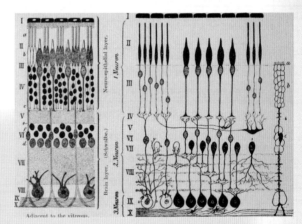

"眼睛"的问题

眼睛有着独特的设计，用来调节焦距、控制采光量，及修正球面像差和色差。如果认为这些都是物竞天择所致，我坦白承认，这个观点荒谬至极。（达尔文：《物种起源》，页 178。）

达尔文很谦虚，他在 150 年前就意识到眼睛之复杂"荒谬"地推翻他的理论。

"The Eye" Problem

To suppose that the eye with all its inimitable contrivances for adjusting the focus to different distances, admitting different amounts of light, and correction of spherical and chromatic aberration, could have been formed by natural selection, seems I freely confess, absurd in the highest degree. (Charles Darwin: *The Origin of Species*, P.178)

Humility of Darwin as he realizes the eye's known complexity even 150 years ago, works against his theories, in an "absurd" way.

9. But They Look So Much Like Us!

Some people get somewhat defensive when I laugh at the scientific data relating apes to humans. They sound very hurt when I say that we are likely not descended from apes, and they often say, "but they look so much like us."

I love to suggest that your dog looks much more like you, than any ape or monkey or chimpanzee, or any of our so-called relatives. Your dog loves you, welcomes you when you come home, and basically gives you a hug and a smile, usually even more than humans. And he knows when you are sick and sad, and tries to comfort you. And when he is sick, your heart breaks, you want to bring him to the best doctor in town, and you are more than willing to pay the best prices for his care, because you feel that he is really *family* and so much like you. I have not known anyone give that much attention to any ape or other so-called ape relatives, except the rare professional ape-handler.

In fact, what I know about apes or monkeys or chimpanzees or any of these so-called relatives is that you *really do not want* any of them in your home. In fact some apes or monkeys or relatives are notorious for being nasty animals, thieves, or destroyers of your home. Just look at the number of cartoons that depict these creatures in a pretty negative manner (remember the apes that supposedly trashed your luggage when you fly commercial airlines). Definitely much worse than depictions of

9. 但它们和我们这么像！

当我嘲笑一些把猿类和人类连上关系的科学数据的时候，有些人觉得有点被冒犯。当我说我们很可能不是猿类的后代时，他们看来很伤心，也经常说："但它们和我们这么像。"

我反而会说你的狗看起来比任何猿、猴子、猩猩，或任何我们的所谓亲属更像你。你的狗爱你，在你回家的时候欢迎你，也会给你拥抱和微笑，有时甚至比人类做得更多。它知道你什么时候生病和伤心，还试图安慰你。当它生病时，你会心碎，你想带它去看城里最好的兽医，你非常愿意为它支付昂贵的医疗费用，因为你觉得它真的是你的家人，还这么像你。我从未见过任何人倾注这么多注意力给任何猿和其他所谓猿类亲属，除了少数专业的猿类照顾者。

事实上，据我所知，不管是猿、猴子、猩猩，或其他任何我们的所谓亲属，你真的**不会**愿意它们住在你家里。其实很多猿、猴子或其他亲属都很令人讨厌，会偷东西，还会毁坏你的家，

显然是我们最好的朋友。
Clearly our best friend.

当你的行李到达机场，破烂如此图时，有些人立刻会想："哪只猴子/猿干的好事？"

When your luggage arrives at the airport, beaten up like this, some people instantly think, "What monkey / ape did this?"

any dogs, who are nearly always described as cuddly, friendly, or even protective of you.

It is quite amazing that everyone has their own favorite animal, and the common excuse given for that choice is something like "he looks just like me." Cats are proud and haughty, but are smart to curl at key moments into your bosom. Rabbits are great to watch as they scamper around and munch, munch, munch away. Horses are truly sensitive, affectionate, and very loyal. But no one that I know would love to have an ape or any of his relatives in their homes.

My favorite animal is actually the otter, even though I cannot actually raise one in my home. I just love to watch otters lie on their backs on the surface of the water, basking in the sunshine, with a large sea conch on their chest. They have perfected the system of feasting on great sea food, by cracking open a sea conch with a rock, beating on the conch as it lies calmly on their chest, and then sucking up the delicious insides.

9. 但它们和我们这么像！
9. But They Look So Much Like Us!

臭名昭著。看看很多动画片都把它们塑造成很负面的形象（还记得当你坐飞机时，把你的行李当垃圾乱扔的猴子吗），肯定比任何关于狗的描述都要糟。人们总是把狗描述为可爱的、友好的，甚至可以保护你。

很有趣的是，每个人都有自己喜欢的动物，选择背后很常见的理由是"它看起来和我很像"。猫是自负和傲慢的，但也会聪明地在适当时机卷缩在你的怀里。兔子很耐看，因为它们跑来跑去又不停咀嚼、咀嚼、咀嚼。马是很敏感、深情而又忠诚的。但据我所知，没有人喜欢在家里养猿或它的其他亲属。

我最喜欢的动物其实是水獭，尽管我不能真的在家里养一只。我就是喜欢看水獭躺在水面上晒太阳取暖，胸前还抱着一个大海螺。它们最懂得吃海鲜大餐，它们会用石头撬开海螺，把海螺稳稳地放在自己胸前敲打它，然后吮吸里面的美味。我甚至想像自己飘浮在海面上看着书，吃这些美味的海鲜。这就是我的想法，这种动物和我很像——喜欢海鲜的书呆子。

诚然，你可以说世界上每种生物都和人类有一定的相似，因为我们有同一位起初的造物主，他根据他所订立的基本原则来创造，又使万物有自己的多样性。所以当你看到**任何动物**，你会说"它看起来和我很像"，这一点也不奇怪。外观上的相似很美好，也给我们提供了鼓励和乐趣，但并不能由此说明一种动物能否或如何不知怎么的神奇地变成另一种（除非是在故事书里）。

I can even imagine myself reading a book on my chest as I lie floating on the ocean eating terrific seafood. That's my idea of an animal just like me, the seafood loving nerd me.

Frankly you can say that every living thing in this world has some resemblance to humans, because there is a common original Creator who created everything in all their diversity, based on basic principles derived from their one Creator. So, it is not really any surprise to find that *any animal* you look at, you could say "he looks just like me." Similarities in outward appearance are wonderful, and provide us with great encouragement and enjoyment, but they do not say anything about if or how one animal somehow magically changes to become another different one (except in story books).

In addition, you can teach an ape, or a parrot for that matter, for even hundreds of years, to try to learn to speak like a human, but they *will never do so*, simply because they are truly different. It is not a matter of education, no matter how much effort and time is put into it. It is fun to do this exercise, but it bears no relationship to whether any animal can spontaneously change to be another one and start talking like us (again, except in story books).

Humans write poetry and story books; but no animals ever do that, in spite of fantasies about how some sounds seem to have poetic rhythm. Humans play violin recitals; apes can never play the violin like a human, even if we try and try to make them. Humans, even so-called cavemen, draw creative pictures on a wall, which uses brain power not that different from the brain power necessary for drawing on a piece of paper, or on a computer. Not even the most intelligent ape can do anything similar. Some animals such as elephants (who are really smart) can be taught to do repetitive rhythmic color designs, but they *cannot actually create*

9. 但它们和我们这么像！
9. But They Look So Much Like Us!

另外，你可以试图教猿（或者鹦鹉）像人一样说话，甚至持续数百年，但它们**永远做不到**，只因为它们真的不同。这不是教育的问题，不管你投入多少时间或精力。做这种实验很有趣，但是这与一种动物是否能够自发变成另一种、并像我们一样说话无关（还是除非在故事书里）。

人类会写诗和故事书；但从来没有动物这样做，尽管有人会幻想某些动物叫声似乎有诗的节奏。人类会演奏小提琴；猿类永远不能像人一样演奏，尽管我们不断试图让它们这样做。人类（即使是所谓原始人）会在墙上画出富有创意的图画，所用的脑力和在纸上或电脑上画画没有什么区别；即使是最聪明的猿类也不能做出类似的事情。有些动物例如大象（它们真的很聪明）可以学会做出有重复规律的色彩设计，但是它们**并不能真的创造独特的艺术作品**，即使是让它们在椰子树下放松，给它们很多鼓励也不能。

人类和所有其他动物的区别就像是白天和黑夜之别，尽管一代又一代人一直试图说这只是层次上的区别。不是的，先生，这**不是层次上的问题**，而是本质上的问题。我们是独一无二的，**独一无二意味着只有一种**。这意味着没有动物真的和我们很像。因为我们是照着上主的形象造的，所以是特别的，很特别的。

翻译：Zhenling Liu

unique art work, in spite of relaxing under a coconut tree and lots of encouragement.

Humans and all other animals are basically the difference between day and night, even though for generations people have been trying to say that they are just a difference in gradations. No sir, it's *not a matter of gradation*, but a matter of total quality. We are unique. *Unique means one of a kind*. That means there are no animals really like us. Because we are made in the image of God, and therefore special. Special.

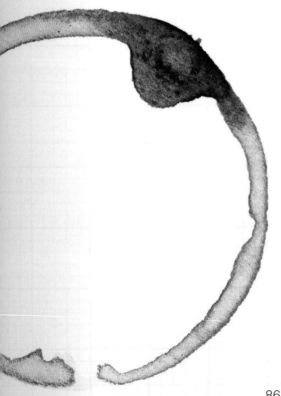

9. 但它们和我们这么像！
9. But They Look So Much Like Us!

和我最像的动物？可能我会想做一只水獭？（享受海鲜的书呆子）
The animal that is most like me? Or I would like to be an otter? (Nerd enjoying seafood)

10. Jubilant Mama Ape Story

The whole ape village came out in great celebration. A new *super ape* had just been born to the most beautiful female ape. All the apes were dancing, jumping and whooping it up. Mama ape was just beaming with joy. Nothing like this had ever happened in this village, or anywhere else.

Suddenly mama ape frowned. "Wait, HOW can I find a future husband for my baby?" Any thoughtful mother would think of that. The music stopped. Everyone stared at her in disbelief. She continued, "Obviously, no one in this village is qualified for my super ape precious girl! Obviously, I cannot just go to the neighboring village to find him!" She started wailing and wailing, and no one could comfort her.

两只猿：如果它们相隔十亿年或相距数万里，我们**怎样**让它们在一起呢？

Two apes: *How* do we get them together, if they are separated by billions of years or thousands of miles?

10. 欢腾雀跃的猿妈妈

整个村子的猿都出来庆祝了。最漂亮的猿妈妈刚生了一只**超级猿**宝宝。所有的猿都跳舞欢呼。猿妈妈只是微笑着,心里充满喜悦。这样的事情在这个村里或是其他任何地方都从未发生过。

突然,猿妈妈皱了下眉头。"等一下,我**怎样**能给我女儿找到未来的丈夫呢?"任何一位考虑周详的母亲都会想到这样的问题。音乐戛然而止。大家都疑惑地看着她。她继续说道:"显然,这个村里没有谁配得上我的超级猿宝贝女儿!显然,我也做不到去邻村找我的女婿!"说着说着,她痛哭起来,大家都无法安慰她。

其他猿开始讨论这个两难问题。"她不可能跋涉千万里,翻越高山湖海,去为她女儿找合适的**超级猿**做丈夫。但是超级猿出生的机会只有百万分之一,更可能只有十亿分之一,所以她势必只好穿越千里去寻找!"这种事**怎么**会发生?

另一只聪明的猿叫喊道:"或者她不得不等十亿年、一亿年、千万年、百万年。但即使是十万年、一万年、五千年、一千年,诸如此类的时间,这孩子都不可能活到那一天!那些

The other apes started discussing this dilemma. "She cannot travel thousands and thousands of miles, over mountains, rivers, seas and oceans, to find the right *super ape* husband for her baby. The chances of this happening are one in a million, or more likely one in a billion, so definitely she would have to go far far away to find him!" *How* was it going to happen?

Another smart ape exclaimed, "Or she may have to wait 1 billion years, 100 million years, 10 million years, 1 million years; but even if it was a hundred thousand years, 10,000 years, 5,000 years, 1,000 years, any number of years like that, it will be impossible for this baby to live that long! It only happens once in billions of chances, the learned ones have told us, so *how* can she wait?"

Let's just go through the possibilities. If perchance, a one in a billion super ape was born, and somehow, in one in a billion chance, got married, and if *magically* there was a super super ape from this marriage, and then, *magically*, there were all these *successively more and more advanced* generations of super super apes, super super super apes, super super super super apes, super super super super super apes, and on and on, I could logically think that the streets of New York or Cincinnati or Beijing would be *swarming* with all kinds of variations of super apes, or super super, or super super super, or...... you get the idea, all superior forms of apes. I should be able to welcome them every day, and say hello, shalom, salam, hola, or even some superior ape grunts to them, with enthusiasm, on all kinds of street corners, all over the world.

I wonder where did they all disappear to? The *lame excuse* is that "oh, they all died off." *Conveniently, not even one* of these maybe thousands of generations and potentially millions or billions of these exotic creatures are ever seen on *any* street, they have all conveniently died off.

10. 欢腾雀跃的猿妈妈
10. Jubilant Mama Ape Story

有学问的说,发生这种事的机会是数十亿分之一,所以她**怎么能等呢**?"

让我们看看可能性吧。如果偶然地,每十亿只猿中有一只超级猿出生,并且不知怎么的,十亿分之一的机会中它们结婚了,并**神奇地**生下了一只超超猿,然后**神奇地**出现了一代又一代**成功进化**的超超猿、超超超猿、超超超超猿、超超超超超猿,等等等等。照理来说,纽约、辛辛那提或北京的大街小巷都会**挤满**各种各样的超猿、超超猿、超超超猿……你明白我的意思,就是各种高等猿类。这样我就可以每天欢迎它们,热情地用各国的语言,甚至用一些高等猿类叽里咕噜的语言跟它们打招呼,遍布街角,遍布世界。

我很想知道它们都消失到哪里去了?有个**勉强的借口**是"哦,它们全都死了"。**太方便了**,这种延续了几千代、可能为数成万上亿的奇异生物在**任何地方都看不到,连一只也没有**,它们都死了,真方便。

并且,**更方便的是**,出于一些不得而知的奇怪原因,尽管挖掘化石已经挖了几百年,都无法找到这各个世代的无数的超级猿,我们仅仅在分散的各地找到几个可疑的残片。玛丽·李奇夫人写了一本坦白得令人震惊的书,揭穿了她的名人丈夫、世界最著名的古生物学家路易斯。

在《往事揭秘》一书中,她写到丈夫雄心壮志地寻找**特定的**奇怪生物化石,当地的发掘工人会**刚刚好**找到路易斯认为"正确"的化石残片交给他。从严格的科学研究角度看,我们称之

And, *conveniently enough*, for some strange unknown reason, even though there have been hundreds of years of digging up fossils, millions and millions of various generations of super apes have all been impossible to find, and we find only a questionable few fragments here and there. In a frank and shocking book, Mrs. Mary Leakey wrote an exposé of her very famous husband, Louis, the world's most famous paleontologist.

In *Disclosing the Past*, she wrote that native diggers would *conveniently* find the "right kind" of fossil fragments to feed to her husband, in his grand ambition to find *just one* of these strange fossilized creatures. In critical science we call that "serious selection bias," and if deliberately done, it is *fraud*.

But, regarding all these selective data point fragments, there is even a quote in *Newsweek*, in the title article, "The Search for Adam & Eve" (1988) that "all the *good* fossils of Africa can be placed in *the palm of your hand*." The palm of your hand! How large is that? Maybe your palm is larger than mine. There probably is some strange excuse for that also.

I have been involved in clinical investigations for decades, so for me, I would like to see some *real statistics on thousands* or tens of thousands of whole person data points and not just fragments, so that we can do some real statistical analyses. In the medical scientific world I live in, we often call the common kind of fossil fragment reporting, as "anecdotal" or "speculative."

In normal scientific investigations in humans, we require hundreds or thousands of individuals to demonstrate any satisfactory *proof*. Drug companies are fond of saying that it takes millions of dollars, or more recently billions of dollars, just to produce one good pill for treatment of cancer, mainly because of the rigor and intensity of collecting uncontaminated data and non-biased selection samples, using thousands

10. 欢腾雀跃的猿妈妈
10. Jubilant Mama Ape Story

为"严重的选择偏差";如果是刻意的,则是**欺诈**。

但是,尽管这些零碎的数据点是经过选择的,《新闻周刊》主题文章"搜寻亚当和夏娃"(1988)仍引用到"所有非洲的**好化石可以全部放在你的手掌里**"。你的手掌!那才有多大呢?也许你的手掌比我的大。这其中也许又有些奇怪的借口。

我从事临床研究几十年,所以对我来说,我须要看到从**成千上万的完整人体数据点获得的真正统计数据**,而不是只有零碎的数据,我们才可以做真正的统计分析。在我从事医学研究的世界里,我们称这类常见的化石残片报告为"趣闻轶事"或"投机取巧"。

请给我真实数据、验证和统计分析。并且不要给我这些想像出来的毛发、皮肤、肌肉和**艺术性的**面部重塑,它只不过**看起来**像是艺术家希望我们想像的样子(你可以从距离辛辛那提机场几分钟路程的地方看到这艺术想像的全息图)。

Please give me some real data, verification, and statistical analyses. And spare me the imaginary hair, skin, muscles and facial *artistry* encasing the fragments, that make it *appear* to be whatever the artist wants us to imagine it to look like (you can see this artful holographed imagination just minutes from the Cincinnati Airport).

of human subjects, comprehensive verification, and precise scientific analyses. Regrettably, these fossil fragment reports are far, far, off from the mark of any rigorous science today.

The key to nearly all science is demonstration of HOW, and the mechanisms involved in the *how*, but fossils are by definition *fossilized*, and therefore *how is presumably impossible*.

All we can say is that these attempts are mostly in the realm of imaginations, fantasies, mythology, or at best philosophy, but please don't tell me that this is really "science," much less proven science. However, mythology does persist and can be very strong in any society, because that's how our lives are lived, based every day on our prevailing assumptions, trends, and *hidden biases*! It would be great to answer the question of that poor mama ape. HOW? She hit the nail on the head!

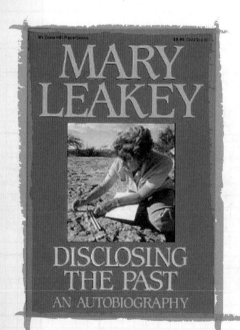

著名的玛丽·李奇揭露了真实的过去：她著名的古生物学家丈夫路易斯利用经过选择、有偏差又零碎的骨头数据，编造了一个支离破碎的故事。

Famed Mary Leakey discloses the real past on selective, biased, fragmentary dead bone data that her most famous paleontologist husband Louis used to weave a fragmentary story.

10. 欢腾雀跃的猿妈妈
10. Jubilant Mama Ape Story

在正常的人体科学研究中，我们需要数以千计的个体数据，以获得令人满意的**证据**。药物公司喜欢说，生产一颗抗癌药物就要花费数百万、近年甚至要数十亿美元。这主要是由于须要收集大量缜密严格的、无污染的数据和无选择偏差的样本，使用数千例受试者，全面验证，以及精密的科学分析。很遗憾，这些化石残片报告与现今的严谨科学大相径庭。

几乎所有科学的关键都是展示"**怎样发生**"及其相关的机理，但是，化石顾名思义就是已经**石化**了，所以**怎样发生就不得而知了**。

我们只可以说，这些尝试大多属于想像、幻想、神话，或顶多是哲学的范畴，但是不要告诉我这是真正的"科学"，更称不上是已被证实的科学。不过，神话的确存在，而且在任何社会中都可以很强大，因为我们每天的生活就是建基于我们压倒性的假设、倾向，和**隐藏的偏见**！我们最好能回答可怜的猿妈妈的问题。**怎样做到呢**？她提出了最关键的问题！

翻译：Grace Lee

11. An Appendix is not an Appendix

The word appendix seems to mean something that isn't that important, just dangling there like an appendage, literally "an appendix."

Growing up, it seems like there were lots of bad stories about the poor appendix. In Asian cultures it is common to *scare* children not to jump around after meals, because the food might drop into that lonely appendix, which is somehow dangling around doing nothing; some rice might get into the appendix and cause appendicitis. This was, and maybe even is, really a great way to keep kids quiet.

As a child, when I jumped around after meals, sometimes indeed I *felt* some pain in my right lower tummy. Since I had a very medical family, and heard many stories about "exciting" medical diseases, I began to imagine that I

倒立对你的阑尾有益吗？
Does standing on the head help your appendix?

11. 阑尾不是附属品

阑尾看起来并不那么重要，只是在那里晃来晃去的，像个附属品。在英文中，阑尾和附属品是同一个单词（appendix）。

我在小时候就听说过很多关于可怜的阑尾的坏话。在亚洲文化中，父母经常**吓唬**孩子，不让孩子在餐后跳来跳去，因为食物有可能掉到那条孤单的阑尾里。那条莫名奇妙、晃来晃去、无所事事的阑尾，一些饭粒掉进去可能就会引起阑尾炎。这可能是，甚至真是一个让孩子保持安静的好方法。

小时候，当我在餐后跳来跳去，有时确实**感到**右下腹有点疼痛。因为我出自医学世家，所以听过许多"令人兴奋"的医学疾病故事，我开始想像我正在**渐渐形成**阑尾炎。

因为我父亲是外科医生，我也梦想成为外科医生，我开始想出不同方法去处理这个"发展中的阑尾炎"。我将头着地倒立，并且拍打我阑尾所在的位置（我**知道**阑尾的位置，因为家里有许多外科的书籍）。我想像那些饭粒或者其他食物颗粒能够从我可怜的阑尾里掉**出来**，这样就可以"制止"我的阑尾炎了。

我想可能是我为着这样的隐隐作痛抱怨得太多了，某个夏天，我发现自己躺在医院的病床上，有真正的医生和护士围着

was *gradually developing* appendicitis.

Since my father was a surgeon, and I had dreams of being a surgeon, I began to devise different ways of managing this "developing appendicitis." I would stand on my head and tap the place where my appendix should be (I *knew* where, since I had many surgery books at home). I imagined that any rice or other food particles would be able to drop back *out* of the poor appendix, thus "preventing" my appendicitis.

At some point, I think maybe I complained too much about this nagging feeling. I found myself one summer lying in a hospital bed, surrounded by real doctors and nurses. And then my father proceeded to remove my appendix. He claimed afterwards that there were signs of "chronic appendicitis" on the pathology report, which may have been diagnosed to confirm my "diagnosis." I understand nowadays it is *not* a real disease and I learned later that doctors also imagine diseases.

One of the college students in my Bible study group SC was from Malaysia; she described how her appendix had ruptured many years ago, and her mother had concluded that if she went to the "moneymaking" hospital, the surgeons might operate in order to make more money. So she decided that it would be more sensible to go to the government hospital, since the government doctors were not paid more money for more operations. Obviously she was a thoughtful Asian mother. Indeed, when our brave girl was brought to the government hospital, even though there was a *ruptured* appendix, the final decision was *not* to go to surgery, since the body's defenses had wrapped themselves around the appendix, and it seemed stable. Today's doctors may or may not agree about that, but SC is a living vibrant testimony to that decision. As she stands straight playing the cello, I can imagine her specially wrapped appendix snugly protected and enjoying the music.

11. 阑尾不是附属品
11. An Appendix is not an Appendix

我。后来,我父亲为我切除了阑尾。父亲事后宣布病理报告显示我有"慢性阑尾炎"的征兆,这个诊断证实了我自己的"诊断"。现在我知道这并**不是**真正的疾病,后来我发现医生也会想像疾病。

我的查经小组有一位来自马来西亚的大学生 SC,她讲述在多年前她的阑尾穿孔,她的母亲认为如果 SC 去了"盈利"的医院,外科医生有可能为了赚钱而动手术。因此,她决定 SC 应该明智地去政府医院,因为政府的医生是不会多做手术就多赚钱的。显然,她是一位深思熟虑的亚洲妈妈。事实上,当这位勇敢的女孩被带到政府医院的时候,即使阑尾**已经穿孔**了,最终的治疗方案是**不做**手术,因为身体的防御机能已经自行对阑尾进行了包裹,病情看来是稳定的。今天的医生可能同意或者不同意以上情况,但是 SC 就是那个决定的活生生的真实见证。当她笔直地坐在那里演奏大提琴的时候,我能想像她那条被特别包裹起来、备受呵护的阑尾也正在享受着音乐。

我在小时候特别想成为外科医生传教士(看我其他的故事!),所以在我的医疗培训阶段,我真的十分高兴能做许多阑尾切除手术。切除阑尾是一件有趣的事,因为这种手术相对"简单",能够满足外科医生去切切缝缝的劲头。无论哪天,给我一条阑尾吧!

但是,有些传教士受困于丛林最深处,那里没有热情的外科医生,无论谁得了阑尾炎,都只能等待,等待身体的防御机制去包裹阑尾,令人惊奇的是他们活下来了。这好比是天然的

When I was a child I had specifically dreamed of becoming a surgeon-missionary (see my other stories!), so in my medical training period I really was quite *excited* to be able to remove many appendices. Appendices are a fun thing to remove, because the operation is relatively "simple," and it satisfies the surgeon's drive to cut and cut, stitch and stitch. Give me an appendix any day!

But indeed there are missionaries stuck in deepest jungles, without the benefit of an eager surgeon, who develop appendicitis, and who can only just wait it out until the body's defense mechanisms wrap around the appendix, and amazingly they survive. These are like natural "scientific" studies, since likely no modern surgeon would dare to *not* operate on an appendicitis patient to test the theory!

Other missionaries have reported how they operated on *themselves*, in deepest jungles, reading from a medical textbook or receiving instructions from some surgeon radioing in the instructions from an outpost outside the jungle. These kinds of stories were part of the great jungle lore for me as I was growing up, dreaming to be a surgeon-missionary.

Just to complete the medical education, please don't think that the appendix is only in the right lower part of the tummy. Unfortunately, the appendix curls around, so sometimes it's in some totally different part of the tummy. Right upper near the liver, left upper near the spleen, left lower on the other side. So your appendix pain may not be just in the usual spot. And the appendix doesn't always point *downwards*; it can point sideways or upwards! So if you are thinking of standing upside down to "prevent appendicitis," (just kidding), you could also remember it isn't really that simple: which way is up or down?

Wasn't that a nice *tour de appendice*? You probably learned more

11. 阑尾不是附属品
11. An Appendix is not an Appendix

"科学"研究，因为应该没有现代的外科医生敢**不**为阑尾炎患者做手术，去证实这个理论。

有另一些传教士曾报告他们是怎样给**他们自己**做手术的。他们在丛林的最深处，参阅医学教科书，或是收听一些在丛林外前哨站的外科医生通过无线电传来的指令说明。这些故事对我来说构成了伟大的丛林传说，伴随着梦想成为外科医生传教士的我成长。

为了完整的医学教育，让我告诉你：不要认为阑尾只在右下腹。不幸的是，阑尾卷曲着，有时在腹腔里的不同部位——右上腹靠近肝脏，左上腹靠近脾脏，或对侧的左下腹。所以你阑尾的疼痛可能并不在通常的位置。并且阑尾也不总是**尖端向下**，可以向两侧或向上方！所以如果你认为倒立能够"预防阑尾炎"（只是个笑话），你也要记住这并不是真的那么简单：哪个方向是向上或向下呢？

这一次**环遊阑尾之旅**精彩吗？你可能学到了很多关于阑尾的知识，甚至比你想知道的还多。然而，如果你在一饱美食之后跳来跳去，感到右下腹疼痛，你可以随时写信给我，我们可以通过无线电、WhatsApp、微信、脸书或电邮沟通，或者你简单一点，去找一位冷静而心灵手巧的外科医生。

确实，阑尾是一个有趣的器官，不是附属品。余下残留物的器官，请看下一个故事。

翻译：Daisy Wang

about the appendix than you ever wanted to know. However, if you feel the pain in your right lower tummy as you are jumping around after a good meal, you can always write me and we can communicate by radio, whatsapp, wechat, facebook, email, or you can simply find a calm surgeon with steady hands and steady brain.

    Indeed, the appendix is an interesting organ, not really just an "appendix". A leftover vestigial organ, read the next chapter.

11. 阑尾不是附属品
11. An Appendix is not an Appendix

在暑期圣经班的传教士故事中扮演医生：你能为你自己切除阑尾吗？
Playing doctor during VBS missionary story: can you remove your own appendix?

12. My Tonsils are Vestigial?

Vestigial means something that is left over somehow from the past. There was at a time, not too long ago, when tonsils and appendixes were considered as "second class organs." " Take them out" was a common surgeon's cry; "they are of no value anyway." At that time, many organs in the body such as the appendix and tonsils were considered "evolutionary vestiges," or something which may have been useful in our "evolutionary past," but which is of "no evolutionary value now." Relegated to second class status.

This kind of concept, of course, is logically considered nonsense by today's standards, since basically everything in the body, even though at first considered useless, seems to be always found to be "valuable" later, just awaiting our belated "discovery."

My own appendix was removed in childhood (see my "An Appendix is not an Appendix" story) for little reason, and undoubtedly affected by this "vestigial organ" concept. In fact I know many people who had operations of the abdomen, who would get their appendix removed, "as a bonus procedure" by the kind surgeon. I wonder how many fine healthy appendices have been removed because of this mistaken belief. And who knows the number of possible complications from people that get their appendixes removed unnecessarily for skimpy reasons.

As another example, doctors used to think that tonsils were "evolutionary vestiges" also, so that there were huge numbers of children who had their tonsils removed on the slightest pretext, such as larger

12. 我的扁桃体是残留的？

残留的意思是从远古不知怎么的遗存下来。扁桃体和阑尾在不久以前仍被视为"二等器官"。外科医师通常会说："拿掉吧，反正它们毫无用处。"在那个年代，体内很多器官像阑尾和扁桃体被视为"进化的残留痕迹"，它们只在我们"过去的进化"中有用，"现在没有进化价值了"，因此被贬到二等地位。

用现在的标准看来，这类观点实在是荒谬，因为身体中任何东西尽管起初看来无用处，后来发现全都是有价值的，只是在等待我们迟来的"发现"。

我自己的阑尾在我儿时因为小小的缘由就被切掉了（参看我的故事"阑尾不是附属品"），毫无疑问是受了"残留器官"这个观点的影响。事实上，我知道很多腹部动过手术的人，他们的阑尾被好心的外科医生顺便切掉，"当作手术的赠品"。谁知道在这个错误观念下有多少好端端的阑尾给拿掉了。谁又知道无缘无故被切掉阑尾的人面临了多少可能的并发症。

另一个例子，医生以前认为扁桃体也是"进化的残留痕迹"，所以大批小孩因为小小的缘由（诸如扁桃体增大或受感

那宝贵的扁桃体在哪里？请再告诉我。
Where is that valuable tonsil, again?

tonsils or infection, resulting in many unnecessary deaths. Nowadays we realize that tonsils are useful for early defense against germs, and that surgery is usually unnecessary, so removal of tonsils has been sharply reduced. The consequent sharp reduction in mortality is now well proven.

But, in a less enlightened time, my wife's brother died as a young man from a "standard" (now considered unnecessary) tonsil operation, and thus I have always had a keen personal interest in this. There was little reason for his surgery, only that it was commonly done for repeated infections of the tonsils.

I have personally seen many children get their tonsils removed unnecessarily. Like many others, I was once a young pediatric intern, pressed into service to check the physical condition of infants and children about to go into tonsil surgery. During these screenings, one could be forgiven for thinking it was a factory process, such was the volume in many hospitals. It was indeed a great revenue generator for hospitals and

12. 我的扁桃体是残留的？
12. My Tonsils are Vestigial?

染）就被切除了扁桃体，结果导致许多本来可以避免的死亡。现在我们认识到扁桃体在早期抵御病菌的过程中是有用的，而那些手术通常是不必要的。扁桃体切除手术的次数因此骤减，随之骤降的死亡率就是明证。

但是在那个不太开明的年代，我的妻舅在年轻时就死于"标准的"（现在认为不必要的）扁桃体摘除手术，所以我对这项手术总是非常关注。他没什么理由要做这个手术，只不过是因为扁桃体反复受感染通常是这样处理。

我亲眼见过很多小孩被不必要地摘除扁桃体。当年我和很多年轻儿科实习医生一样，也曾被迫去为将要接受扁桃体手术的婴儿和儿童做身体检查。在筛查过程中，即使感到这是工厂化作业也无可厚非，因为在很多医院都有大量的此类手术。实际上，它也无意中成了医院和医生的庞大收入来源。

不时有年轻人死于扁桃体手术后的伤口出血，由于出血位置在咽后壁的隐蔽部位，发现时已经太迟了。"进化的残留痕迹"这个观念可谓悲剧。

过去，医生又认为松果体是进化的残留痕迹，但是现在我们认识到它和许多其他腺体一样，是重要的内分泌器官，掌控着人体的基础节律（身体中任何器官组织都有节律，尽管我们至今尚不完全了解）。

举例来说，松果体分泌的一种相对较新发现的激素——褪黑激素，很多人服食来控制国际飞行的"时差不适"。我们得到的教训是，既然造物主的设计是那么精准，身体中任何器官

doctors, just incidentally.

Young people die at times from the tonsil operation because bleeding after surgery from the tonsil area may not be recognized until it is too late, since the bleeding occurs in a hidden area at the back of the throat. Belief in such concepts as "evolutionary vestige" can be tragic.

In the past, doctors thought that the pineal gland was an evolutionary vestige, but now we know that it is an important endocrine organ that controls basic body rhythms (everything in the body has rhythms which are not totally understood as yet), among many other things.

For example, one of its relatively newly discovered hormones, melatonin, has been popularly used to control "jet lag" when we fly internationally. The lesson is that we should not easily dismiss organs in the body as "useless," since the Creator's designs are truly ingenious, and unlikely to be "vestigial." The term "vestigial" itself is now vestigial, as you can see from the following story.

One day I was giving a lecture on a newly discovered hormone at the time, calcitonin, which comes from the thyroid gland (you can read my "That Funny Thyroid" article in Reggietales.org). It was exciting to measure that new hormone in babies, something which few people had done. So one learned professor asked me a leading question after my lecture: "do you think this is a vestigial hormone?" implying that the hormone really might not be of much value in "modern life".

I jokingly replied, "when was the last time you saw a vestigial hormone?" implying that hormones considered "vestigial" years ago have been discovered later to have value, often having an amazing function that we in our ignorance have not yet "discovered" or imagined. The assembly of top scientists roared in laughter, catching the joke and agreeing with the gist of my riposte. Laughter is a good way to poke fun at

12. 我的扁桃体是残留的？
12. My Tonsils are Vestigial?

都不会是"残留的"，我们不应该轻易把它们当做"无用的"就切掉。从下面的故事你会发现，"残留的"这个词汇本身现在已经是残留的了。

有一次，我发表关于当时新发现的激素——降钙素的报告，此激素来源于甲状腺（你可以到 Reggietales.org 阅读我的文章"有趣的甲状腺"）。我很高兴可以为婴儿检测这种新激素的水平，这是先前没有人做过的。我报告完毕，有位博学的教授问了我一个很有诱导性的问题："你认为这是一种（进化后）残留的激素吗？"暗示这种激素在"现代生活"中其实可能没有什么价值。

我开玩笑地回答："你最后一次看到残留的激素是什么时候？"暗示多年前被视为"残留"的那些激素，后来发现都有价值，通常是因为我们自己无知，才未曾"发现"或料想不到其神奇功能。一众顶尖的科学家哄堂大笑，他们领悟了这个笑话，同意我的机敏回答。笑声是对科学上的狂妄自大最好的嘲弄。

又有一次，在美国位列前茅的辛辛那提儿童医院，泌尿科教授发表报告。他为膀胱构造十分异常的孩子设计了一种新型的膀胱。他惊人地把孩子本身的阑尾移植到膀胱，使其成为膀胱的出口通道，这是令人振奋的突破性手术。这件事对于外科医生和我们来说，确实都需要一些想像力（想想用阑尾作为膀胱引流的通道），但这真是一项完美的重建。

scientific pomposity.

One day at the Children's Hospital of Cincinnati, which is a foremost children's hospitals in the USA, the Professor of Urology was speaking. He had designed a new urinary bladder for children with very difficult bladder anatomy. He took the appendix of the child, and startlingly moved it to become the outlet channel for the bladder, which was a fascinating breakthrough operation. This really takes some imagination, both for the surgeon and for us (imagine a funnel like bladder draining into the appendix), but it was a perfect reconstruction.

I happened to be chairman of that particular meeting, a "grand rounds," where the entire medical staff come together weekly for a clinical presentation. I made a concluding comment, "now I know why God made the appendix," accompanied by echoing laughter. It is indeed my goal as a scientist, to find out why God made this or that, in this or that way, though not necessarily by making natural structures do unusual functions (that is why God made surgeons!!)

12. 我的扁桃体是残留的？
12. My Tonsils are Vestigial?

我恰好是那个特别会议的主席，那是所有医生都参加的每周临床病例讨论"大会"。我总结说："我总算明白了为什么上主会创造阑尾。"大家在笑声中共鸣。身为科学家，我的目标正是探索为什么上主会创造这个或那个，为什么造成这样子或那样子，却又有时破例让自然的构造出现不寻常的表现（这就是上主创造外科医生的缘由！）。

翻译：Sonic

这位未来工程师在将来也许可以投身人体部位重组的工作。
Budding engineer could find a future role in re-arranging body parts also.

13. I Love Salmon

I moved from Ohio to the Northwest, where salmon are famously plentiful. I have always loved to eat salmon, especially grilled salmon. So, it is a perfect place for me to think of this great fish. I especially appreciate the fact that, with salmon, there is plenty of meat to bite into, inexpensively, and I can eat it to my heart's delight. This is in contrast to the sophisticated tender fish that I can eat in Asia, but which is so expensive, and is served in extremely *small* portions even at sumptuous banquets, that I can barely satisfy my fish hunger! Come to Seattle and taste the juicy **thick chunks** of salmon. It's literally like eating a sizzling Argentinian beef steak, and I am sure it is much healthier!

各种鲑鱼，不同的生产和包装，一样美味。
Salmon of all kinds, available in many forms, all delicious.

13. 鲑鱼情结

我从俄亥俄州搬到因盛产鲑鱼而闻名的美国西北地区。我一向喜欢吃鲑鱼,特别是烤鲑鱼,所以我来到这地方就自然会想到它们。鲑鱼非常多肉,价格还合理,每每吃到它,我都会发自内心感觉到无比满足。这与我在亚洲吃过的精致细嫩的鱼截然不同:那些鱼价格不菲,即使在豪华宴会上也以极**少**的份量装盘,根本不能满足我吃鱼的渴望。欢迎来到西雅图,品尝多汁又**大块**的鲑鱼,它完全可以媲美炙热的阿根廷牛排,而且更加有益健康!

简单而丰富的鲑鱼餐,让我垂涎。
A simple but ample salmon meal, just makes me salivate.

其实我还非常欣赏鲑鱼的生平故事,这是一个值得我们学习的神秘历险故事。事实上,当它们在西北地区溪流里奋力前进的时候,我可以现场观察到他们生命之旅的一些片段。你也许知道鲑鱼通常须要**逆流而上一千英里**,同时湍急的流水将它们推往**相反**的方向。它们还要在遇到巨大的石头和其他障碍物

What I really like also is the life story of the salmon fish. It is a story of adventure and mystery, that we might even learn from, and I can actually watch part of its life journeys in real time, as the salmon swim heroically in the streams of the Northwest. You might know that the salmon often need to swim a *thousand miles upstream*, against the rushing waters pushing them in the *opposite* direction. And, they have to physically jump up at times to go over huge rocks and other obstacles, several *feet* high. This journey is so strenuous that watching them makes me feel tired. What amazing drive must they have to do this! And how do they know that they *have to do this*? What really is the inner drive?

Close to and facing my Seattle home, there is actually a river that salmon have to use to swim upstream. But because of the four-lane road construction through my local town, the government had to spend millions of dollars to make sure that the road did not cut off the stream, and block or interfere with the salmon run. In fact, the state of Washington built and rebuilt many roads like that, running into billions of dollars, which is a reflection of how serious Washingtonians view any interruption of this *magic salmon run*, so it is not a joking matter!

And then in some amazing way, the super marathon swimmers can find the original little stream they traveled from, and even the original spot where they were born and grew up, a long time, maybe years, ago. You can imagine the many bifurcations and forks in the streams as they thrash their way upstream, where they have to make instant decisions as to which path to take. Seattle land traffic is considered one of the worst in the country, because the roads wind around lakes and hills, up and down hills, zipping sharply left and right, and changing names all the time; the road that is my back road changes names 6 times in 15 minutes of driving! It's all very unnerving, and I'm nearly always using a GPS,

13. 鲑鱼情结
13. I Love Salmon

时,用力向上跳跃几**英尺**高。这个旅程是多么的费劲,我光是看着就已经感到非常疲惫。它们一定有强大的内在动力才会这样努力吧!它们为何会认定自己**必须这样做**呢?它们的真正驱动力是什么呢?

在我西雅图家的对面有一条河,正正是鲑鱼逆流而上的必经之路。它位于城市扩建四车道的工地附近。政府为了不干扰鲑鱼洄游,不得不花费数百万美元来确保河流没有被切断或堵塞。而事实上,在华盛顿州还有许多类似的道路重建或改建工程,涉及数十亿美元。这正好反映了华盛顿州居民的认真态度:**神奇的鲑鱼洄游**不可被阻断,这可不是开玩笑的!

这些超级马拉松游泳健将以神奇的方式,找到它们几年前游经的河道,甚至是自己很久以前(数以年计)出生、长大的地点。你可以想像河流中有许多分岔和支流,在它们逆流而上的瞬间,必须快速决定游往哪个方向。西雅图地面交通被认为是全国最糟糕的地方之一,因为这些道路环绕着湖泊和丘陵,

英勇地逆流而上。逆流一千英里。

The heroic upstream swim. 1,000 miles upstream.

Global Positioning System, to find my way. What sort of high powered GPS does the salmon have? And there are not even names on the streams. A wrong turn and they *never* get to their original home of birth. And can you imagine, they do this in a *fasted* state, drawing on their fat and other reserves in the body. Just imagine yourself driving while hunger pangs are gripping you; I have, and it isn't funny, and that's why I always have a nut bar in my car glove compartment, just in case.

As the salmon return to their home, they mate, at this literal and poetic high point of their life, in the beautiful snowcapped mountains of the Northwest. Their life journey is finally over. Having spent all their energy making this long strenuous journey, they die, literally exhausted, essentially *sacrificing themselves* for a new generation to come. They have come back to their ancestral home, sweet home to fulfill their destiny, a fitting location after their many trials of life. I can't but think, that, Seattle being where my mother was born, grew up, and went to university, before she took the "slow boat to China," I am literally "coming home," also. Back to where my mother grew up, among the white snow covered tall pines and firs, next to that huge Pacific Ocean, in the lovely State of Washington. In my life, I have had the pleasure of traveling thousands of miles, and eaten lots of good sea food dinners: I cannot say my life has been exhausting, and I have never had any significant fasting times (work in Asia comes with too much great food!). Yet in the turns and jumps over the years, I could *poetically* empathize with my fishy friends. But I have come home, finally, likely to live my last years here!

The newborn salmon, hatched in this gorgeous environment, wait for the spring flowers in full blossom, and then some salmon species decide, en masse, together with hordes of other very young salmon, to go! "Where we going?" could be the legitimate question that one young

13. 鲑鱼情结
13. I Love Salmon

不断上山下山,还有大量"之"字形的设计。而且相连的道路可能会用不同的名字。以我家背后的一条路为例,在大约 15 分钟的车程中,它的名字就变了六次!这令人感到非常不安,所以我总是依靠全球定位系统(GPS)来找路。鲑鱼是用哪种高效能 GPS 的呢?那些河道甚至没有名字啊。一次错误的转弯,它们就**永远不能**回到它们的家乡。而且它们在**禁食**状态下洄游,单靠身体中的脂肪和其他储备上路。我明白饿着肚子驾驶的痛苦,所以我一定会在车子的储物箱里准备一条坚果营养棒,以备不时之需。

当鲑鱼回到家乡,它们在西北部白雪皑皑的山顶交配,达到它们生命(事实上及诗意上)的高峰,它们的生命之旅也终于结束。它们花费了所有的精力,完成了漫长而艰辛的旅程,油尽灯枯而死,根本是为了下一代而**牺牲自己**。在经历了无数的生活体验之后,它们回到甜蜜的老家,以实现它们命中注定的终结。我不得不想到:我的母亲漂洋过海去中国之前,就是在西雅图出生、成长和读大学。回到西雅图,于我而言也是"回家",回到可爱的华盛顿州,在太平洋边上,在白雪覆盖的高大松柏当中,是我母亲长大的地方。在我的人生中,我已经遊历了数千英里,尝遍了大量的海鲜美食。当然,我的生活并没有令我枯竭,我也并不曾捱过饥饿(甚至在亚洲工作期间享受了大量美食!)。然而,在多年来的辗转和奔波之后,我可以**诗意地**代入我的鱼类好友的处境:我终于回家了,并会在此度过余生!

salmon asks the other. "I don't think we know," might be one answer. Or "I'm not sure, maybe the others know." "In any case, we *have to go*, so let's go, we'll soon find out!" And so, a new generation of tens of thousands fly down the streams. Other salmon species decide to take their sweet time growing up, and enjoying themselves, in the many mountain streams and lakes, for even a year, before finally taking their plunge downstream. I'm a baby doctor (neonatologist), so I like to imagine that, many of the early swimmers are really just my kind of *neonatal* babies, and yet they can furiously swim swiftly downstream with great purpose: where did they get the intelligence to do all that? how could they do that while so very young? And with *no parental guidance*, no PG, as Americans like to say about movies, no one to advise them at all, about the long and dangerous journey ahead of them.

Down, down, indeed they go, downstream, the thousand miles their parents had struggled in the *opposite* direction, until they reach the spots where the stream waters mix with salty water of the great Pacific Ocean. Are they in for a shock? I wonder if they might be asking one another again, "Where are we?" "What's going on here?" "The water here feels very different." Indeed, their bodies now have to adjust to the new very different salty environment: how do they do that? But, their body begins to adapt, and on and on they still must go, until they realize now they have hit the real gigantic salty Pacific Ocean!

From now on, how do they know where they should be swimming? There is a pathway *ordained* for them in the vast ocean, that is not marked for human eyes to see, but which they know, *instinctively*, what it will be, along a way that their ancestors used to swim. These elliptical pathways in the ocean can last from 18 months to 8 years to traverse. And each specific path is only for each specific species of salmon. Coho, pink,

13. 鲑鱼情结
13. I Love Salmon

鲑鱼宝宝在这美丽的环境里出生，并在此等待春天的花蕾绽放，然后有一些品种的年轻鲑鱼将会决定联群结队去遊历！"我们去哪里？"顺理成章成为新一代鲑鱼之间的话题。它们的答案可能是"我不认为我们知道。"或者"我不肯定，也许有其他鱼知道。""无论如何，我们**必须走**，所以我们出发吧，我们会很快找到答案的！"所以，新一代无数的鲑鱼飞快地向下游进发。另一些品种的鲑鱼决定先留下来，幸福地成长，在众多山溪和湖泊中好好享受，甚至长达一年，然后才顺流而下。我是儿科医生（新生儿科），所以我会想像，那些很早就出发的小鱼就是我的那些**新生**婴儿，然而它们怀着伟大的目的，就勇敢地、快速地展开生命征程。它们做这一切所需的智慧从何而来？它们是那么的幼小，**没有父母的指导**，没有"家长引引"（电影分级制度的用词），没有任何人给予建议，它们怎么面对漫长而危险的前路呢？

它们确实出发了，顺流而下，朝着与它们父母挣扎了一千英里的**相反**方向，直至它们抵达河水流入太平洋的混合水域。它们会震惊吗？我猜想它们是否又再互相询问："我们在哪里？""现在发生什么事？""这里的水似乎很不同。"确实，它们的身体现在必须适应截然不同的咸水环境，它们怎样做得到？然而，它们的身体不得不开始逐渐适应，直到它们意识到自己已经完全沉浸在浩瀚太平洋的咸水中！

从现在开始，它们怎么知道自己应该往哪里游？在广阔无垠的海洋中，鲑鱼在**命中注定**的航道上迈进。这航道不是人眼

chinook, sockeye salmon, each have their *own pathway*, and they do not mix their paths. And, to complicate matters, "Americans don't mix with Asians": the Northwestern salmon certainly do not mix paths with Asian salmon coming in their own pathways from the *other* side of the Pacific Ocean, in their own long circles to return to the *Asian* continent, even if their paths, from East and West, may even seem to cross in the middle of the Pacific.

These great ocean pathways could each run 10,000 miles, and yet the salmon, now maturing as they swim, somehow know that is their destiny. I can imagine my inquisitive mother, on the boat to China, in the early 1930s, having a chance to chat with the captain of the boat about the complexity of the routing that he would have to take all the way, through wind and storm and waves, using relatively sophisticated compasses, radios, and stars to safely land thousands of miles away. How do the salmon manage their ocean navigations? Many mariners would love to

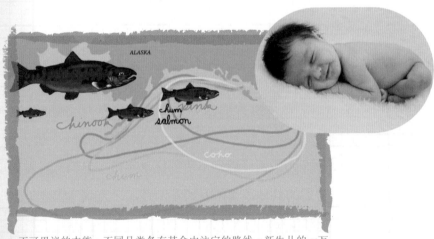

不可思议的本能，不同品类各有其命中注定的路线。新生儿的一万英里长征。（由西雅图艺术家 Lily Heinzen 绘画）

Impossible instinct, to each their own ordained path. Newborns on 10,000 mile odyssey. (Drawing by artist Lily Heinzen, Seattle)

13. 鲑鱼情结
13. I Love Salmon

所能看到的，但是它们靠着**本能**就知道，就是他们的祖先曾经游过的路线。沿着海洋中的这些椭圆形洄游通道，鲑鱼须要游上 18 个月甚至 8 年之久，而每个特定品种的鲑鱼都有其特定的路线。银鲑、粉红鲑、帝王鲑、红鲑，每种鱼都有**自己的洄游通道**，它们之间不会混淆。而且更复杂的是，"美洲的与亚洲的不相往来"：西北地区鲑鱼的路线绝对不会与太平洋**彼岸**的亚洲鲑鱼混淆，而对方也最终会返回**亚洲**，即使东、西双方漫长的洄游通道会在太平洋的中间交错。

这些漫长的海洋通道动辄上万英里，然而不知何故，在游泳当中日渐成熟的鲑鱼似乎认定这就是它们的宿命。我想像，我那爱探究的母亲，在 1930 年代初坐船去中国的时候，有机会与船长聊天，谈到他是如何运用复杂而精准的罗盘、无线电和星象的位置，在经历了暴风雨和巨浪的洗礼之后，安全抵达几千英里外的彼岸。鲑鱼是靠着什么航海的呢？水手们一定很想知道吧？也许他们可以**有所学习**，从而改善自己的航海技术呢。

这巨大的椭圆形洄游通道最终回到了海洋之旅的起点，接着这些鲑鱼就像它们的父母以前一样逆流而上，开始另一次**英勇的、周而复始**的伟大壮举。当鲑鱼回到溪涧和河流时，有些被捕捉，成就了我餐桌上的美味。甚至在写这篇文章的时候，我仍在回味着鲑鱼的滋味。

鲑鱼的生命旅途，令科学家困惑了几百年，也持续让我感到惊讶。我身为科学家，也对这个巨大的谜题发出惊叹，而它

know; maybe they could *learn something* to improve their own navigation.

This huge elliptical ocean circle pathway ultimately comes right back to where they began their ocean journey, after which the salmon then go upstream as their parents once did, for another *heroic cycle* of great feat. As they go up the creeks and streams, some are captured, and a few land on my dinner table, to my great delight. I can even savor their taste this minute as I'm writing this.

But what has puzzled scientists for hundreds of years, continues to fill me with amazement. As a scientist myself, I stand in awe of this huge puzzle, actually just one among thousands of other spectacular puzzles in nature. The Designer who was involved in designing every single minute detail of a salmon's life has created such an interesting enigmatic life journey story, even for the lowly salmon for which I have a special love. And I am grateful for that, for biology and for my taste buds.

Biologically speaking, we lump these puzzles as puzzles of *instinct*. Darwin literally "gave up" on instinct, as impossible puzzles of nature that upset his basic views of life, and his theories of how we came to be. I agree, and the really funny thing is, we are finding, with time, that there are now *more and more impossible* puzzles, not less, the more we learn from biology, from the largest giants of nature, right down to the tiniest particles of life. Instinctive behavior seems programmed and designed after all, and no "natural mutations," or "natural selection" could ever produce that.

13. 鲑鱼情结
13. I Love Salmon

只是自然界无数惊人的谜题之一。鲑鱼虽然微不足道，但是我特别钟爱。为了生物学，也为了我自己的饮食喜好，我非常感激设计鲑鱼每一个生活细节的那位设计师，创造了一个有趣又神秘的生命旅程故事。

在生物学上，我们把这些谜题统称为**本能**之谜。达尔文根本上"放弃"了本能，因为这个自然界不可思议的谜题，摧毁了他对于生命的基本观点，以及人类从何而来的理论。有趣的是，随着时间的推移，我们在生物学上知道得越多（大至自然界中的巨无霸，小至最微小的生命粒子），**不可思议**的谜题反而**越来越多**，没有减少。毕竟，本能行为似乎是事先设定和设计好的，根本不可能是"天然突变"或"物竞天择"的结果。

翻译：Anya Zhang

许多本能都是那么不可思议，以致它们的发展在读者看来，大概是一个足以推翻我全部理论的难点。我在这里先要声明一点，就是我不准备讨论智力的起源，就如我未曾讨论生命本身的起源一样。我们所要讨论的，只是同纲动物中本能及其他精神能力的多样性。

达尔文的哀叹，出自他的经典著作《物种起源》

Many Instincts are so wonderful that their development will probably appear to the reader a difficulty sufficient to overthrow my whole theory. I may here premise that I have nothing to do with the origin of the mental powers, any more than I have with that of life itself. We are concerned only with the diversities of instinct and of the other mental faculties in animals of the same class.

Darwin's lament, from Darwin's original classic, *Origin of Species*.

14. Genius is Genius: The Genius of Genius

Through life we have at times encountered geniuses, or heard about geniuses in history. With modern media, we often can witness geniuses, in action on YouTube or Facebook, or described in popular magazines. But have you ever wondered where geniuses come from? How come they are geniuses?

You could say, "I know a genius when I see one." And that's not bad, because when you see a genius, ideally you should instantly recognize that his or her abilities are quantum leaps from anyone else's, and he/she should not be just a graduated small difference from an accomplished person, of which there are many. For example, if you see a child conduct a symphony flawlessly, we immediately know, "that's a genius." If we can't see that, then maybe it is still not at genius level?

I am not talking about high IQ only, or an Olympic champion, or the most famous movie star. In fact, when Mensans who are in the Mensa Club of highest 2% IQs are asked if they consider themselves geniuses, they all demurred. In fact, it has been famously difficult to really classify on paper what is a genius, so the "I know when I see" one definition is still pretty good. For the analytic minded I suggest that geniuses *at least* have to have the following 4 components. I have modified this definition, from a book I once read on genius, which I remembered well, but cannot re-locate: 1) layered complexity, 2) harmonic interrelationships, 3) elegant

14. 天才就是天才：天才中的天才

我们在生活中有时会遇到天才，或者听说过历史中的天才。在YouTube、脸书等现代媒体上，或者流行杂志中，我们也可以见到活生生的天才。但是你可有想过天才从何而来？为什么他们是天才？

你会说："天才，我看见就知道。"那也不错，因为当你看到一个天才的时候，你会立即发现他或她的能力和其他人有天壤之别，和那些芸芸大众中稍有成就的人相比，他或她不只

当我亲眼看见，就一定知道那是天才。

I can certainly know a genius when I see one in action.

unity, 4) inspirational or poetic clarity. I think that's not a bad list, *as a start*, which helps us analytically to classify and help guess, if we need to, that we are seeing a genius before us.

1) A genius has to be able to handle certain extremely complex matters easily, like an orchestra conductor or composer who knows *many layers of complexity* at the same time, precisely and logically. 2) Genius includes *harmony*, like all the members of the team in an extremely complex surgical operation, who have to be in harmony with each other, because, if not, the extreme surgery will fail. 3) In genius, all components and functions together have to have *unity in an elegant way*, otherwise

波音 747 对比最简单的细胞

- 4.5 x 10^6（百万）个零件 = 波音 747
- 1 x 10^9（十亿）个零件 = 最简单的细胞

Boeing 747 vs Simplest Cell

4.5 X10^6 (6th power, million) parts=Boeing 747
1 X 10^9 (9th power, billion) parts=Simplest Cell

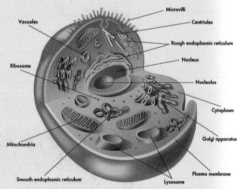

天才的品质，以所谓"简简单单的细胞"为例，就反映出上主无尽的天赋。
Qualities of genius, example of so called "simple cell," reflection of God's infinite genius.

14. 天才就是天才：天才中的天才
14. Genius is Genius: The Genius of Genius

是一星半点的区别。举例来说，当你看到一个小孩完美无瑕地指挥一曲交响乐，我们立即知道："那是天才"。如果我们当时没有看出来，也可能就是未及天才的程度吧？

我所说的不是单单高智商，或奥运冠军、最著名的电影明星之类。事实上，只接受智商最高的2%人参加的门萨会，他们的成员被问及他们是否认为自己是天才时，他们全都否定。其实我们都知道很难清楚界定什么是天才，因此"我看见就知道"这个标准也是相当不错的。如果要分析，我觉得天才**至少**得有下面四个特质。我以前读过一本关于天才的书，我修订了书中的定义。这本书我记得很清楚，但是现在找不到了。这四个特征分别是：1）多层次的复杂，2）和谐的相互关系，3）优雅的统一，4）灵感或诗意的清晰呈现。我觉得这个清单不错，可以**作为起点**，帮助我们分析归类和必要时去猜测我们面前的天才。

1）天才能够同时了解**复杂事物的多个层面**，处理极端复杂的问题时既轻松、精确又合乎逻辑，就像管弦乐团的指挥或者作曲家。2）天才具备**和谐**，就像一台极端复杂的外科手术，团队中所有成员之间都要和谐，因为不和谐这个大手术就会失

天才的特质

1) 多层次的复杂
2) 和谐的相互关系
3) 优雅的统一
4) 诗意（灵感）的清晰呈现

Qualities of Genius

1) Layered complexity
2) Harmonic interrelationships
3) Elegant unity
4) Poetic (inspirational) clarity

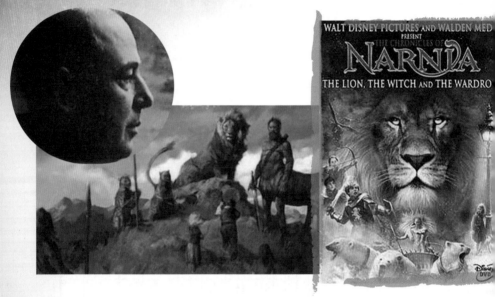

我最喜爱的作家 C.S. 路易斯（《纳尼亚传奇》）是天才的例子，他拥有所有的天才特质。
My favorite author, CS Lewis (*Chronicles of Narnia*), example of genius, in all its qualities.

even if it is layered complexity plus harmony in interrelationships, without an elegant unifying theme or direction, it still would not be super-outstanding. 4) And to be really genius, there has to be a special qualitative **inspirational or poetic brilliance** that comes through in **perfect clarity**, the awesome ahhh factor, something like only one in 10,000 people could "pull off."

For example, spectacular writers like Shakespeare are able to pull the components of their masterpieces together with layered complexity, harmonic interrelationships, elegant unity, and inspirational clarity. That's why we call Shakespeare a genius, or someone like him. Or just use the simplest definition of "I know one when I see one." It really is probably just as good.

14. 天才就是天才：天才中的天才
14. Genius is Genius: The Genius of Genius

败。3）作为天才，所有的组成部分和功能都要以**优雅的方式统一起来**，否则，就算有多层的复杂、和谐的相互关系，没有优雅的统一主题和方向，也很难出类拔萃。4）要做真正的天才，还要拥有特殊的**灵感或诗意的光芒**，以**完美清晰**的方式呈现出来，要万中无一，令人拍案叫绝。

好像莎士比亚这样的出色作家，他们可以通过多层的复杂、和谐的相互关系、优雅的统一，和灵感的清晰呈现，把他们的杰作的各个部分连接起来。因此我们把莎士比亚这一类人称为天才。也可以用最简单的定义："我看见就知道"，那也是不错的。

不管天才的定义到底如何，你应该从生物学的角度问一个更重要的问题："天才从何而来？"事实上，科学家在天才起源上也遇到了障碍。其实根本就没有答案，即使有些科学家自己都被视为天才！没有人相信（至少在他们心底深处）特定的猿类会**真**的进化为特定的人类天才。一些人也许会说进化论**一般而言**可以解释每件事，但是当他们要坦诚面对这个问题，例如简单地被问到"怎样发生？"，请他们解释这个进程的每一步如何发生，他们都被难倒了。

猿不能变成天才，不管它多机灵，就是这样简单。千万年过去了，时间对于进化为天才有何助益？没有，花上千万年的时间都不能做到。"怎样发生？"这个简单的问题再次难住了所有人。没有人可以说天才源于**突变**，一个突变把普通人的基因变为天才基因，或者无数个突变把一个"泛泛之辈"变为天

So, no matter what the definition really is of a genius, you might really ask the even more important question, from a biologic view, where does genius come from? In fact, scientists have come to a total roadblock in understanding the origin of genius. In fact, there is no explanation at all, even though some of the scientists might themselves be considered geniuses! Nobody believes, at least deep down in their hearts, that a specific ape could *really* evolve into a specific human genius; some might say that evolution *generically* solves everything, but when faced with this problem frankly, like being asked a simple question of "how?", meaning please explain every step of "how" this works, everyone is stumped.

An ape could not become a genius, no matter how smart, it's that simple. And millions and millions of years, how does that help to evolve a genius? It doesn't, and there's no way that millions and millions of years would conceivably do that. Again, the simple question, "how?" floors everyone. Nor can anyone explain the origin of genius as being a *mutant*, a sudden mutation that transforms an ordinary person's genes into genius genes; or zillions of sudden mutations that transform an "ordinary person" into genius; there is not a single gene that has been found to move one through mutations into a genius, in spite of now knowing hundreds of thousands of genes and their mutations. Mutations are mutations, and no one has ever associated a genius with mutation. To call a genius a *mutant* is a total contradiction. Or an insult. "How?" is the scientific question, and it is here that the logic stops. This impasse results in a *deafening silence* among scientists when they realize they are totally stumped.

In point of fact, geniuses can *only* come from our ancestors' genes, and if you are a genius, the more pristine your ancestor genes were, the more likely was he or she also a genius, or was carrying your genius genes. Your genius genes are undoubtedly ancestral (there is no other

14. 天才就是天才：天才中的天才
14. Genius is Genius: The Genius of Genius

才。尽管我们现在已经认识了成千上万个基因及其突变，但是仍没有发现哪怕一个基因会通过突变把人变成天才。突变就是突变，没有人会把突变和天才联系在一起。把天才视为**突变是完全矛盾的**，甚至是侮辱。"怎样发生？"是个科学的问题，也是他们的逻辑止步之处。那些科学家意识到他们被难到了，这个僵局带来**震耳欲聋的沉默**。

事实上，天才**只可能**源于我们祖先的基因。如果你是天才，而你祖先的基因更为纯粹，那么他或她就更有可能是天才，或者携带着你的天才基因。你的天才基因毫无疑问是祖传的（你越想会越肯定没有其他选择），一代代传下来，而这些基因越能保持纯粹，没有污染，没有参杂，没有突变，就越能**完好无缺**地传递给你。因此，如果你是天才，你之所以能成为天才，唯一合乎逻辑的原因是你的祖先就是天才，或者这些基因在他们身上隐藏着被保护起来，直到合适的时间和机会才展现出来，于是你收获了这些基因的成果，光芒四射。你的天才基因肯定不是源于丛林里的猿经历千万年进化，也不是基因突变！

因此，把这个逻辑更推进一步，可能我们很多祖先都拥有天才基因，经年以后，这些基因或许被稀释、被污染、被参杂，或发生突变，因此我们**不太是**天才了。又有可能，我们远古祖先的基因更纯粹、更古老，所以即使只是他们当中的"普罗大众"，也拥有比我们"更好、更聪明的基因"。惊讶吧？我们仔细想想就会明白，在可观察的科学和自然界中，你其实只能

option, when you think more about it), coming down through each generation, and the more pristine these genes could maintain themselves, without polluting effects, contaminating effects, or mutating effects, the more likely were your genius genes transmitted *intact* to you. So, if you are a genius, the only logical way you can be a genius is because your ancestors were geniuses, or had these genes hidden and well protected, until the right time and opportunity to be revealed, and you are reaping the brilliance of their genes. Definitely your genius genes have not come from some jungle ape, over millions and millions of years, nor a mutant!

So, to push this logic even further, it is possible to consider that many of our ancestors had genius genes in them, which over the years has likely been diluted, polluted, contaminated, and mutated, so that we have become *less* than geniuses. And it is possible that, even in the "general population," the more pristine and the more ancient the gene had been, our early and ancient ancestors likely had "better and smarter genes," than ours. Surprise!? When we think more about it, in observable science, in nature, you can really only go from more perfect to less perfect, consistent with the pesky universal 2nd law of thermodynamics pushing events downhill.

And this is why when we examine ancient history, and hear of people who designed the pyramids, or the Inca architectures, our first reaction might be, how could they do this? Aren't they supposed to be dumb, and we are the evolved superior humans? In point of fact, the reason they could do it, is that they were likely really *smart*, *really*, *really*, *smart*, meaning they could be like geniuses! And they could do these amazing things *without* any IBM Watson so-called smartest machines, or any of 10s of thousands of advanced technologies accumulated over many generations that we now have! The myth that we are "smarter" than our

14. 天才就是天才：天才中的天才
14. Genius is Genius: The Genius of Genius

从比较完美走向不太完美，与那讨厌却普及的热力学第二定律正好一致，都是推动事件走下坡。

这就是为什么当我们研究古代历史，听到当时的人建造了金字塔或印加古建筑，我们的第一反应就是：他们是怎样做到的？难道他们不是笨笨的吗？我们才是进化了的高等人类。事实上，他们能够做到这些事情，是因为他们**很聪明，真的真的很聪明**，即是说他们就像天才一样！而且他们做这些奇妙事情的时候，**没有 IBM 沃森之类的所谓高智能机器**，也没有我们经过许多代人所积累下来的成千上万的高级科技！我们误以为我们比祖先更"聪明"，只不过在于我们今天拥有那么多的科技。哪怕一天，甚至一个小时没有手机，没有任何现代装备，我们会怎么样？至于印加人怎么能修建马丘比丘古城，埃及人怎么能修建金字塔，还会有其他原因吗？（我们不必去幻想虚构的外星人。）就只是因为他们拥有比较纯粹的基因，这些基因很可能比我们所估计的更为完美，也很可能比我们现代人所拥有的"更好"。

中国老话说：**一代不如一代**，是有道理的。想想现今我们每个人有多少突变。根据粗略估计，每个人有 15 至 30 个潜在的重大突变，而且一直增加，没有减少。随着代代相传，新的**突变悄然侵入**，又有一些赖着不走。因此，抱歉，我们祖先的基因很好。我猜想，实际上他们的突变率比较低，大概是因为他们较少污染、较少参杂、较少辐射，经过很多代之后，这些因素令我们的基因变得没有原来的好了。如果说进化论是神，

ancestors, is mostly just a matter of technology, which we have so much of today. Where would we be, even for one day, maybe even one hour, without our cellphones, and all our modern gadgets? As to how Incas could build Machu Picchu and Egyptians build pyramids, what other reason would there be? There is no fantasy need to invoke imaginary extra-terrestrial aliens. They had simply more pristine genes, that were likely more perfect than the ones we credit them with, and likely "better" than the ones we have now.

There is an old Chinese saying, that each generation is not as good as the previous generation. "*yi dai bu ru yi dai*" (literally, one generation, not like, meaning less than, the *previous* generation). The ancients knew what they were talking about. Just think about how many mutations we each have today. A ballpark estimate is each person has 15 to 30 potentially significant mutations. And they keep increasing, not decreasing, as generations multiply, since new mutations start *slipping in*, and some hang around. So, I'm sorry, but our ancestors had it good. I'm guessing, realistically, that their mutation rate was less since, presumably they had less pollution, less contamination, less radiation, all of which over the generations probably hasn't made our genes any better. I am an "atheist" regarding the god of evolution, and favor The Creative Genius God, since ultimately, even before the original pristine genius genes, must be *the first, the original, the truly pristine, the greatest Genius, who created genius.*

14. 天才就是天才：天才中的天才
14. Genius is Genius: The Genius of Genius

我就是一个无神论的人。我赞同有一位充满创意的天才造物主，因为追本溯源，在最原始最纯粹的天才基因之前的，一定是那位**首先的、原本的、真正纯粹的、创造出天才的、最伟大的天才**。

翻译：Sonic

我特别喜欢第四种天才特质——灵感清晰，就像主所创造的美丽花朵，每一朵都表明天才的清晰灵感。

I love especially the 4th component of genius, of inspirational clarity, just like beautiful flowers of creation, each declaring the inspirational clarity of its genius.
(© 2018 Elaine Tsang)

15. Mystery of Language Part A: My Fascination with Language

I have a great fascination with language. My first language was, strangely, English, because of my American born mother, but I learned to also speak Chaozhou, Cantonese, Mandarin, and finally, Hakka (Kejia), in that order. I learned to speak the first 3 in childhood, and the last 2, Mandarin and Hakka, as an adult. I learned Mandarin in the Cincinnati Chinese Church, and Hakka by speaking with my wife's nieces in Seattle, who grew up in Thailand speaking the language!

Since prayer language might include a few "more specialized terms," I used to only pray in English, my "comfortable" language. I'm wondering if there was also a bit of subconscious absurdity that my language should be "better" when addressing the heavens! But I have always urged people to be bold in speaking any language, so I took my *own* advice and started using Mandarin in prayers. At first it was infrequent, only when I really had to, like when I was praying with someone who only spoke Mandarin. In the last year, however, since moving to Seattle, better late than never, I started using Mandarin in prayers regularly, and also Cantonese (which oddly, I had *not* used either before), when praying with my wife, who knows all the languages mentioned, plus childhood Thai.

Then, to my own surprise, and to my wife's even greater surprise, I actually began to try using Chaozhou and Hakka in my prayers. At age 76! Amazingly, I can now converse and even say my prayers, pretty fluently, in

15. 语言的奥秘（上）：
语言令我着迷

我对语言很着迷。虽然我在香港出生，但是因为我母亲生于美国，只会说英语，英语就成了我的母语。不过我先后学会了潮州话、广东话、普通话，还有客家话。孩提时期我就学会了前三种语言，成年以后再学会普通话和客家话。我在辛辛那提华人教会学会普通话，在西雅图与我太太的侄女们聊天当中学会客家话。她们在泰国长大，从小就说客家话！

因为祷告的时候常会有一些"专用词"，我以前只用英语祷告，因为我说得最顺溜，不知是不是我潜意识中荒唐地认为，我跟上天说话的时候就应该用我"最拿手"的语言。但是我总是鼓励别人要大胆地说各种语言，所以我也接纳我自己的建议，开始用普通话祷告。起初我只是在不得已的时候偶尔为之，例如当我和只会说普通话的人一起祷告，就只好用普通话。不过，一年前我搬到西雅图后，想着迟做总比不做好，就开始在和太太一起祷告的时候，经常用普通话或广东话（不知怎么回事，我以前也不曾用广东话祷告）。前面提到的这些语言，我太太全都会说，另外她从孩提时期就会说泰语。

后来，出乎我的意料，更出乎我太太的意料，我甚至开始

all 5 languages! Now, I can even jump between different languages for fun and verbal effect, since I feel it "stretches my mind."

Actually, I think I had a distinct language advantage when growing up, since many languages, and especially these 5 languages, were floating around me then. The words and their nerve tracks obviously had *etched* themselves into my brain speech zones, subconsciously, just waiting for the right time to integrate with my later attempts at speaking them. As I began to actually speak in my less familiar languages, and especially as I prayed, suddenly some words I *did not even* know I knew, just popped into my mouth, sometimes seemingly bypassing my thinking brain, out of some *deep brain reserve*, words I might only have heard 65 years ago! And many related memories have also surprisingly floated into my mind. What a surprise and mystery!

Some people think that Chinese languages are all just *dialects* of Chinese. But language experts now know that many Chinese languages are truly *languages*, and no longer considered dialects. For example, Chaozhou, Cantonese, and Hakka are languages different from Mandarin, as different as Spanish, Portuguese, and Italian are from Latin. Meaning that a person speaking in Mandarin cannot just switch into Chaozhou or Cantonese like a "dialect" or a tone change, because, surprisingly, the actual words used in speaking may *not even be Chinese words* that a Mandarin speaking person would use.

What is confusing to Mandarin speakers is that the Chaozhou or Cantonese person, in trying to communicate with other Chinese people, converts himself into Mandarin automatically when he *writes out his words*. So, when other Chinese look at the words, they recognize it as Mandarin Chinese, and think that the *spoken* form of Chaozhou or Cantonese, which is their *true natural language*, must also be a variation

15. 语言的奥秘（上）：语言令我着迷
15. Mystery of Language Part A: My Fascination with Language

尝试用潮州话和客家话祷告。76岁了！我现在竟然能颇为流利地用这五种语言对话甚至祷告。我甚至可以在这五种语言之间随意转换，我觉得这样能"拓展我的思维"，既好玩，又能加强语意。

回想我的成长经历，其实我有得天独厚的语言优势，因为在我周围能听到很多种语言，尤其是上述这五种。这些言谈及其神经音轨显然已经**深深刻入**我的大脑语言区，在我的潜意识中，只是等待合适的时机，在我日后尝试说的时候就迸出来了。当我开始尝试用这些相对陌生的语言，尤其是在祷告的时候，有些我**从不知道自己晓得的字词**会脱口而出，甚至好像没有经过我大脑的思考，直接就从**脑袋的深层存储区**中跳出来。这些字词可能只是我在65年前听到过的！并且很多相关的记忆也被勾起。真是惊讶！不可思议！

有人认为中国的各种语言只是一些**方言**，但是语言专家现在确认很多都是真正的**语言**，而不仅仅是方言。举例来说，潮洲话、广东话、客家话都是独立的语言，它们与普通话之别，不亚于西班牙语、葡萄牙语、意大利语与拉丁文之别。也就是说，说普通话的人不能像说"方言"那样，只是通过声调变化就说出潮洲话或广东话，因为出乎意料，这些语言中有些字词**在普通话中根本没有**。

令说普通话者困惑的是，当说潮洲话或广东话的人与其他中国人交流的时候，往往会自动翻译成普通话再**写出来**。所以当其他中国人看到这些字，就认定是普通话，然后就认为潮洲

of Mandarin Chinese. It might, and it might not; it just depends.

To further complicate matters, these *spoken* languages might nowadays be also *written out* in their own languages, and not in official Mandarin. Open a truly Cantonese newspaper in Hong Kong, or try to *read the ads* in the Hong Kong MTR subway. The Mandarin speaking person will be so surprised, and suddenly realize that a *different language* is being used! Isn't Hong Kong part of China? Strange, is it not? What are they trying to say, is it Chinese? Well, it is and it isn't.

This situation is an analogy of Europe of the past, when the literate European communicated in *Latin* in writing, but spoke Spanish, Portuguese, or Italian at home or with friends. Got it? As a modern joke, they might now communicate in English writing! Which is obviously not Spanish, Portuguese or Italian.

Try to read the Kanji of the Japanese language, which are basically Chinese characters, but which sound, and might mean different things from Chinese. Chinese is basically often indeed like the Latin of Asia, the structure around which many languages are related, including the so called "Chinese languages," and other national languages (such as spoken Korean, Japanese, Vietnamese).

And just for fun, just within the Chinese languages, remember that there are 9 tones in Cantonese, compared with 4 or 5 in Mandarin, which adds a tremendous complexity and strain on speaker and listener, when trying to learn this as a new language. I always quip that when you are not a native Cantonese speaker, the moment you open your mouth to try to speak in Cantonese, even if you think your version of Cantonese is pretty good, there is an *8 to 1 chance* that you will be wrong!

And to the native speaker, when the non-Cantonese says something that is not the right tone, it is very sharply recognized as hilarious

15. 语言的奥秘（上）：语言令我着迷
15. Mystery of Language Part A: My Fascination with Language

话或广东话的**口语**（对方**真正的自然语言**）肯定是普通话的一种变异。这或许对，或许不对，视情况而定。

更有甚者，如今当地人常常把这些**口语写成文字**，而不是以正规普通话写出来。香港的正宗广东话报纸，或香港铁路上的**广告**，会让说普通话的人摸不着头脑，因为这根本就是**另一种语言**！香港不是中国的一部分吗？奇怪！他们说的是什么？是中文吗？也许是，也许不是。

这种情况和以前的欧洲非常类似。当年有文化的欧洲人用**拉丁文**写作，但是在家里或者朋友之间，还是说西班牙语、葡萄牙语或意大利语。明白了吧？现代人开玩笑说，如今欧洲人大概都靠写英文沟通了！显然，英文不等同于西班牙语、葡萄牙语或意大利语。

读读日语中的汉字，它们基本上是中文字，但是读音不同，字义可能也不同。中文字就像是亚洲的拉丁文，周边很多语言文字都是以之为基础的，包括各种所谓"中国语言"和其他国家的语言（例如韩语、日语、越南语等）。

看看谷歌尝试把这段粤语翻译成英语，却以为它是普通话。其实真正的意思是："我不懂的，就是不懂；考试很具挑战性；我竟然无法回答任何问题；课堂上，我什么都听不懂。"一个香港孩子的慨叹，在社交媒体疯传。

Watch Cantonese translated into English by Google assuming it's Mandarin. Real meaning is more like "what I don't know, I don't know; exams are challenging; I couldn't answer any question; in class I didn't understand anything." A viral lament by a Hong Kong kid.

我最喜欢的满书架自学语言用的教学光碟。
My favorite book shelf of self-teaching language audios.

or weird, just like any music that is *off key*, which it is, sort of. And unfortunately, the native speaker (and therefore listener) will often burst out, embarrassingly, with great laughter, because to him or her it really sounds "so funny." That's why it is so difficult to learn Cantonese if you are not a native speaker! It can be quite humiliating, intentional or not.

I used to play language tapes as I was driving, in order to learn different languages. At my peak of learning, I could speak 20 to 50 phrases, in each different language. My minimum target was "20 phrases in 20 languages." The great linguist Berlitz' recommendation has always been that if you knew 20 phrases in any language you can get around in any country, since the only really important words are "how are you, thank you, goodbye, pleased to meet you, where is the bathroom? etc."

I had great fun learning these languages and truly wherever I traveled, I would use the local language that I memorized, using key

15. 语言的奥秘（上）：语言令我着迷
15. Mystery of Language Part A: My Fascination with Language

更有意思的是，就在中国语言当中，广东话有九个声调，普通话则有四至五个声调，令初学者说或听的时候大大增加了难度。我常开玩笑说，如果你不是从小就说广东话，哪怕你自认广东话说得不错，只要一开口，你的发音**有九分之八的机会**是错的！

而对于母语是广东话的人而言，别人的声调稍有不对，听起来就会很怪异或者很滑稽，就像音乐中听到**走音**一样。听者往往会忍不住捧腹大笑，令对方尴尬，因为走音听起来实在是"十分好笑"。正因为这样，广东话十分难学，会在有意无意中令初学者感觉颜面扫地。

为了学新的语言，我曾经时常边开车边听录音带。在高峰期，我可以用各种语言说 20 到 50 句短语。我给自己定的最低目标是"用 20 种语言说 20 句话"。语言学大师贝立兹建议，只要学会任何语言的 20 句日常用语，你就基本上可以在那个国家玩得转，因为真正重要的只有几句："你好！谢谢！再见！很高兴见到你！洗手间在哪儿？"等等。

我学外语学得很开心，并且无论去到哪里，我都真的会用当地语言说我所记得的关键短句，尤其是泰语、马来印尼语、越南语、日语、韩语、土耳其语、阿拉伯语、希伯来语、印地语，和欧洲各国语言。学外语对我并不难，因为我早就有至少五种语言的音轨刻印在我脑海里，所以于我而言，只是在已知的**声音或音符**上一直加添或改动。用对方的母语说寥寥几句，常常

phrases. Particularly for Thai, Malay-Indonesian, Vietnamese, Japanese, Korean, Turkish, Arabic, Hebrew, Hindi, and European languages. My language acquisition was not difficult because I had already had at least 5 language tracks etched in my brain, so it was just a series of add-ons and modifications of already known *sounds or musical notes*. It was often a great icebreaker just to say a few words in the listener's mother tongue, and it opened many doors. See my Uncle Reggie Stories, "Salam Allay Kuhm."

There is also no question that the younger you learn a language, the more fluent you are. I've always noted that children from China who come to America before the teenage period learn within a year to speak fluently like an American, those who come as teenagers become nearly American, but those who come after age 25 will nearly never speak totally like an American! For example, I visited Thailand for the first time at age 15, and I tried to learn the language, so in later years when I returned to visit, locals often commented that my accents were pretty good, likely since I got an early start.

鹦鹉总归是鹦鹉，再怎么"聪明"，也绝不可能朗诵莎士比亚的作品。

Parrots are parrots, no matter how "smart" they are, no Shakespearean oration ever.

15. 语言的奥秘（上）：语言令我着迷
15. Mystery of Language Part A: My Fascination with Language.

都可以轻松地破冰，从而开启很多的门。请参看我写的曾叔叔故事："Salam Allay Kuhm"。

毫无疑问，学语言是越年轻越容易说得流利。我注意到从中国来的孩子，如果是没到青少年期就来了，往往在一年之内他的英语就可以和美国孩子一样流利；如果是青少年期才来的，他的英语也会说得很接近美国人；但是那些25岁以后来的，几乎完全不可能像美国人那样地道！举个例子，我15岁那年第一次去泰国，当时学了点泰语，多年后我再去泰国，当地人就夸我的发音，估计是因为我学得早的缘故。

尽管学新的语言乐趣多多，并且语言能帮助我们跨越文化障碍，有效地沟通并建立友谊，但是，**谁都不知道世界上各种语言最初从何而来**！无人知晓！人们想尽各种办法尝试发掘语言最初的起源，例如通过教鹦鹉和猿说人话，虽然好玩，但最终还是不能解开谜团。人类与非人类在语言能力上有天壤之别。这可不只是声调或口音的细微差别，而是关乎鱼类与人类之间的生理跃变。

未完待续：语言从何而来？……

翻译：杨迪霞

But with all this great joy of learning different languages, and using language to cross different cultural barriers for effective communication and friendship, *no one knows, in the secular world, where language originally comes from!* No one! People have tried all kinds of ways of trying to find out where language originally came from, by teaching parrots and apes to try to speak some kind of human language, but the results, though fun and interesting, are abysmal in understanding this mystery. And the gap between humans and nonhumans in language ability is gargantuan in scope. We are not just talking about a small little difference or an accent, or a tone difference. We are talking about a huge jump, physiologically, that is, well, between a fish and a man.

Continued in Part B: Where Does Language Come From?...

15. 语言的奥秘（上）：语言令我着迷
15. Mystery of Language Part A: My Fascination with Language.

世界上有 6,000 多种语言，没有人知道语言本来是如何首先"开始"的，除了真的是"上主的作为"。仅在中国、印度、非洲或印度尼西亚群岛，**每个**区域就有大约一千多种语言。

More than 6,000 languages of the world, and no one has a clue how language originally first "began," except truly as an "act of God." Roughly one thousand or more languages *each*, just in China, India, Africa, or the Indonesian Archipelago.

16. Mystery of Language Part B: Where Does Language Come From?

...Continued from Part A: My Fascination with Language

Yes, I am fascinated by language. But where did language come from?? Realistically, it is not even possible to imagine how language or it's "precursors" *could ever* transition, from fish to man, in any way, shape or form. No matter how you twist, distort or shape the "precursors" of the human brain, nerves, tongue, vocal cords and larynx voice box. In spite of the huge understanding of modern DNA, there is not the slightest credible finding of a transitional DNA jump from apes to humans, or from birds to apes, or from any species to the next species, in terms of the fantastic jump that is required along this theoretic "language phylogenetic pathway," in order to "finally" generate real human language, *any* human language.

The complexity is unquestionably mind-boggling. It is already fantastically difficult to even imagine designing a system of making the sounds for *any one word*, let alone the construction of the sounds of words in phrases, in Shakespearean novels, in scientific jargon, in mathematics, in philosophical language.

And to think that there are 6,000 languages at least in the world, implying that the DNA, brain, nerves and muscles etc. etc. have to all have the *potential readiness* to meet this varying language challenge, to be

16. 语言的奥秘（下）：语言从何而来？

……续上篇：语言令我着迷

是的，我迷上了语言。但是语言从何而来？实际上，根本没有人想像得到语言或其"前体"**如何能够**以任何方式、形态或形式从鱼类过渡到人类，无论你如何扭曲、窜改或塑造人类大脑、神经、舌头、声带和喉部发声部位的"前体"。就着这一套"语言演化路径"理论所须要的神奇跃变而言，尽管现代科学家对 DNA 已经非常了解，但是从猿类到人类、从鸟类到猿类、从任何物种到下一个物种的转变，都找不到丝毫可信的过渡性 DNA，可以"最终"生成真正的人类语言，**不论哪一种人类语言**。

语言之复杂无疑令人难以理解。我们极难想像如何设计一个能说**任何一个单词**的发音系统，更不用说能发声讲出短语、莎士比亚小说、科学术语，或数学和哲学用语。

世界上至少有 6,000 种语言，这意味着 DNA、大脑、神经和肌肉等都必须**准备就绪**去迎接不同的语言挑战，才能够说出

able to speak *in any* one or few of the 6,000 plus languages. It is like the "system" is just waiting, at birth, to get ready to activate to meet one or few *specific* oncoming language challenges. "Any" child, who is not deaf or neurologically damaged at birth, can rapidly learn as a child, *any language that he or she is born into*. Isn't that amazing?! "The system" is ready for all these many options, meaning that "the system" instantly knows how to adapt to think and speak even in the rarest birth language.

And this happens, no matter what race or ethnicity the child is, or wherever the child is born, in Africa or near the North Pole. And even the so-called wildest, and most "primitive" human being can speak from early babyhood and childhood. And he or she doesn't just speak a "generic human language," but the *specific* language of his or her tribe, to communicate very well with the other tribal members. And not just that language, but with the *right accent, tones and grammar* that the next tribe over the mountains may or may not be able to understand. A language that *we* do not understand, nor likely ever will, even though we are considered well educated.

Isn't that shocking, and unquestionable, that every tribal child learns his own language, even though he doesn't go to school and he doesn't have a language teacher threatening him with a stick and blackboard. Often, so called primitive languages turn out even to be extremely complicated, and not in the least bit "primitive," whatever that means. And the child definitely doesn't need to learn to use a laptop computer or smartphone, in order to communicate quite effectively, with other tribal members, in the beautiful language of his birth.

And once so-called natives start moving around, and meeting other natives from the next mountain, ergo, you start finding that they can learn, and soon can be *multi-linguistic*, and do it far better than any

16. 语言的奥秘（下）：语言从何而来？
16. Mystery of Language Part B: Where Does Language Come From?

6,000多种语言中的**任何**一种或几种。就好像这个"系统"在刚刚出生时就等待着，准备启动去应付一个或几个迎面而来的、**特定**的语言挑战。"任何"出生时没有耳聋或神经损伤的小孩，都可以迅速地学习**任何他或她的出生环境中的语言**。这不是很神奇吗？"系统"已经为所有这些选项做好了准备，即是说就算在最稀有的语言环境中出生，"系统"也立即懂得如何调整去思考和说话。

无论孩子是什么血统或种族，无论孩子出生在什么地方，在非洲还是在北极附近，都是这样。即使是所谓最野蛮、最"原始"的人类，也可以从婴儿时期和孩提时期的早期就开始说话。他或她不只是说一种"通用的人类语言"，而是说他或她部落的**特定**语言，可以与本部落其他人好好沟通。不仅仅是那种语言，而且会用**正确的口音、声调和语法**，甚至是山上隔壁部落都未必能理解的。这种语言**我们不懂，将来也不会懂**，即使我们算是受过良好教育。

即使没有上学，也没有语言老师用棍棒和黑板来威胁，每个部落里的孩子都学会他自己的语言，这难道不令人震惊，而且不容置疑吗？所谓原始语言通常非常复杂，一点都不"原始"（无论那是什么意思）。孩子绝对无须学习使用手提电脑或智能手机，就能够与其他部落成员用美丽的母语相当有效地沟通。

一旦那些所谓土人开始四处走动，并与隔壁山上的土人会面，你就会发现他们能够学习，很快就**能说多种语言**，并且比任何来自哈佛或耶鲁大学的语言学教授都要好得多。我和泰国

linguistic professor from Harvard or Yale. As I travel with tribal people in the mountains of Thailand, when they meet others, they seem suddenly to be able to communicate across very different tribal languages. As a University educated city person, this often comes across as truly humbling.

My white American missionary friend's son, who grew up in Turkey, can speak in 6 or 7 languages of the region including English, Turkish, Arabic, and Chinese. That's very impressive of course, but many kids I teach in primitive mountain tribe areas of northern Thailand often speak in Thai, Lisu, Akha, Lahu, Chinese and English. When I teach them English, they also seem to pick it up quite well, and especially their pronunciations are often spot on. Frankly, often better than some others I have taught who are considered "more sophisticated," usually from a background of speaking only one language. I imagine it's because tribal kids *already have* so many languages, words and sounds in their brain that they can modify their existing sound tracks and switch to a new language more easily. Just like my childhood multi-linguistic background. The plasticity and *receptivity* of the brain in early life, especially, to learn any language, is really just amazing, and a shock to realize.

And in today's world, isn't it fun to be able to switch from language to language, using language translation apps? I use them all the time. The free Pleco electronic language dictionary app is tops for English/ Chinese conversions, try it. I use these apps immediately when I see difficult texts in non-English news reports, Facebook or WeChat. And any long article can now be Google translated right away. It may not be perfect, but I can easily catch the gist. Or even simply click on the text, and the English translation might come up right away for me, without even "copy and paste." I can imagine that these instant translations can only get better.

16. 语言的奥秘（下）：语言从何而来？
16. Mystery of Language Part B: Where Does Language Come From?

山区部落的人一起旅行的时候发现，当他们见到其他人时，似乎突然间就能够透过不同的部落语言来沟通。身为受过大学教育的城市人，这种情况常常让我学会谦逊。

我的美国白人传教士朋友的儿子在土耳其长大，能说英语、土耳其语、阿拉伯语和华语等六、七种语言。这当然非常令人印象深刻，但是我在泰国北部原始山区部落教导的许多孩子，也通常能说泰语、傈僳族语、阿卡族语、拉祜族语、华语和英语。当我教他们英语时，他们似乎也学得相当好，特别是他们的发音通常是完全正确的。坦白说，他们通常比我教过的其他被认为"更有学识"、来自单一语言背景的人说得更好。我想这是因为在部落孩子的大脑中**已经有了**这么多语言、字词和声音，他们可以较容易修改已有的音轨，切换到新的语言，就像我童年的多语言背景一样。早年大脑的可塑性和**可接受性**（特别是学习任何语言的能力）的确非常奇妙，令人震撼。

在现今世界，使用语言翻译应用程序就能够在语言之间切换，不是很有趣吗？我一直在用。免费的 Pleco 电子词典应用程序是中英文转换的首选，试试吧。当我在非英文的新闻报导、脸书或微信上一看到很难的字词，就会立即用这些应用程序。现在任何长篇文章都可以用谷歌立刻翻译出来。它可能不完美，但可以让我很容易抓住要点。甚至不必"复制和贴上"，只须简单地点击文字，英文翻译就马上出现。我可以想像，这些即时翻译只会变得越来越好。有趣的是，当我写作和翻译这些曾

Funny enough, as I am writing these Uncle Reggie stories, and getting them translated, I come up with words often that Google Translate doesn't do so well, but they "humbly" ask me if I can "offer a better translation," which often I can! So, when you use Google Translate, who knows, it could very well be from our Uncle Reggie team of translators.

As a translator myself at church, usually from Mandarin into English, the greatest challenge is when I have to perform instantaneous headphone-to-microphone translations from the room behind the assembly hall. It is then that I recognize that the brain, nerves and muscles are amazingly able to *instantly* translate, synchronize and coordinate the words flying through the air, wires and brain. Even as the speaker is speaking in Mandarin, my brain has moved simultaneously into English, and out of my mouth comes English, as fast as the speaker is going. I am always astounded that it can really happen just like that, but it does, like at the United Nations I suppose.

But in reality, many bi or multi-cultural people normally function like that in any multicultural and multilinguistic society, translating furiously and quickly all the time between languages. Sometimes just in

谷歌原本的歌翻译。
Original Google translation.

16. 语言的奥秘（下）：语言从何而来？
16. Mystery of Language Part B: Where Does Language Come From?

叔叔故事的时候，我常常碰到一些单词谷歌翻译得不太好，但是他们会"谦虚地"问我能否"提供更好的翻译"，这是我常常可以做到的！所以当你使用谷歌翻译，得出的结果可能是我们曾叔叔翻译团队提供的，谁知道呢。

我自己作为教会翻译员，通常从普通话翻译成英语，最大的挑战是要在礼堂后面的房间里，进行耳机到麦克风的即时翻译。这时候我就认识到大脑、神经和肌肉是那么神奇，能够**即时**翻译、同步化和协调那些在空气、电线和大脑间飞跃的字词。讲员用普通话说话，我的大脑同时转移到英语，口中说出英语，就像讲员一样快。我总是很震讶它真的可以这样，也确实发生了，就好像在联合国那样。

但事实上，在任何多文化、多语言的社会中，许多双文化或多文化的人平常都是这样运作的，一直都是疯狂而快速地进行语言翻译。有时只是在大脑中，有时会说出来。当然，我假设不久之后，我们就能用谷歌眼镜等工具和一个完全说外语的

建议的翻译。我们可以提出建议去改进翻译结果。
Suggested new translation. We can give suggestions to improve the translation.

the brain, sometimes talking it out. And of course, with Google glasses etc., I am assuming that soon we could all be talking to a person that is speaking a totally foreign language, and the words will spill out onto our magic glasses or into our ear pieces, in our favorite language. What a fun time that will be!

People used to question how could the Supreme Intelligence be so intelligent as to be able to hear everyone speaking at the same time to him (like during prayers), in different languages, but now I have some idea of how it might happen. He's wearing some kind of Super Master Translator Eyeglasses connected by Super WIFI to the Greatest Computer System of the Universe, or more likely His brain is already simply *The* Master WiFi Computer Translator, or something like that!

But the really simple reason He could hear and understand all languages is because He created language! That is the only logical way it could have happened, in my view. The great Designer who designed the entire universe, designed humans with the innate and unique ability to communicate in the beauty and mystery of language, any language. I don't think there is any other way it could have happened. In order that we can communicate with other humans and also with the Designer Himself.

In any case, without any doubt, the greatest language that transcends all languages, to me, better than any of the 6000 plus known languages, is the *language of love*. That language was most dramatically, personally delivered, and sealed by Christ: in the ultimate love language of personal sacrifice, a language that can be understood in *any language*. His Book of the Greatest Love is translated into the greatest numbers of languages for any book, and linguists are constantly busy translating it into *every language* of the world, even those with originally *no written language*. It

16. 语言的奥秘（下）：语言从何而来？
16. Mystery of Language Part B: Where Does Language Come From?

人说话，这些话将会以我们最喜欢的语言，涌现于我们的魔术眼镜或者耳机里。这将是多么有趣的时刻啊！

曾经有人质疑，那位至高的智慧怎么可能如此聪明，能够听到每个人用不同语言同时向他讲话（例如在祈祷中），但是现在我大概想到是如何发生的。他戴着通过超级 WiFi 连接到宇宙最伟大电脑系统的超级大师翻译眼镜，或者更有可能的是，他的大脑已经就是**那个**大师 WiFi 电脑翻译机，或者类似的东西！

但是，他能听到并理解所有语言，真正的原因很简单，就是语言是他所创造的！我认为只有这样，这一切才合乎逻辑。这位伟大设计师设计了整个宇宙，他设计的人类天生具备独特的能力，能够用任何美丽而神秘的语言来沟通。我认为没有任何其他方式可以促成这一切。这是为了使我们可以与其他人沟通，也可以与那位设计师自己沟通。

无论如何，对于我来说，**爱的语言**毫无疑问就是最伟大的语言。它超越所有语言，比 6,000 多种已知语言中的任何一种都更好。这种语言已经由基督最淋漓尽致地亲自传达并确认了，他用自己的牺牲表达出终极爱的语言，**任何语言**的人都可以理解。他那部关于最伟大的爱的书，在所有书当中被翻译成最多种语言，语言学家一直忙于把它翻译成世界上的**所有语言**，甚至是那些原本**没有书面语言**的语言。这是可以做到的，因为爱的语言推动着这项工作，真的推动到世界的尽头。

翻译：Hongyan Zhu

can be done, because the language of love drives this work, truly to the ends of the earth.

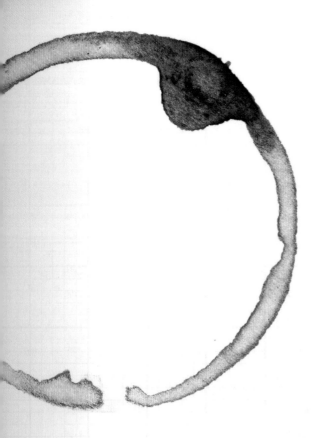

16. 语言的奥秘（下）：语言从何而来？
16. Mystery of Language Part B: Where Does Language Come From?

阿卡族语的赞美诗和傈僳族语的圣经，当中使用的书面语言，由传教士语言学家设计，并教给这些亚洲少数民族部落的人。用**爱的语言**，即使在泰国偏远的部落地区，也能被数百万人理解。

Akha language hymnal and Lisu language Bible, using written languages designed, and taught to Asian minority tribes by missionary linguists. In a *language of love*, understood by millions, even in remote tribal locations of Thailand.

17. Be a Skeptic Part A: Dogmas Come and Go

I've spent many decades of my life doing highly competitive academic medical research, writing lots of papers, obtaining many NIH grants, and speaking at numerous academic conferences literally in every continent. Feel free to Google *Reginald Tsang* (there is strangely only one such name) if you'd like to read more. From all of these experiences, I've learned to be *very skeptical* and even sometimes cynical. Skeptical, primarily because, surprisingly, a lot of so-called wisdom, especially in clinical medicine, routinely gets erased every 10 or 20 years. Even though everyone in the field seemed so initially positive about it, sometimes even claiming unanimous opinion, it turns out there was a basic error in understanding, and, with new thinking, major revisions had to be made. Therefore, skepticism is good practice in science. In fact, *even cynical* can be good, because not everyone is altruistic about research. Desire for fame and fortune could easily seep into a highly competitive arena.

Related to my academic interest, currently there is huge controversy about whether you need vitamin D as a daily supplement, or not. I get this question from friends often, and at scientific meetings in different parts of the world, after my lectures. In the beginning of my career, decades ago, there was a relatively clear-cut definition of vitamin D deficiency, and we knew only a few people actually had it. Then, the definitions for "deficiency" and "insufficiency" were redrawn, mostly based on

17. 做个怀疑者（上）：信条恒变

我在竞争激烈的医学范畴从事了几十年学术研究，写过许多论文，许多次获得美国国家卫生研究院（NIH）的资助，在世界各大洲都做过多次学术报告。如果你还想知道更多，可以在谷歌上搜索我的名字 Reginald Tsang（奇怪谷歌上没有人和我重名）。我的这些经验告诉我，要**秉持怀疑**甚至是猜忌的态度。怀疑，主要是因为临床医学中的许多所谓智慧，竟然都会例行地在 10 至 20 年间被推翻。尽管一开始在学术范畴中大家一致看好，但后来就会发现原来在理解上有个基本错误，新的想法随之而来，于是要作出重大修订。因此，在科学范畴，常存怀疑之心是好的。其实，**甚至是猜忌**也是好的，因为科学研究者并非全是利他的。功名利禄很容易渗透到高度竞争的竞技场当中。

目前人们对于是否需要每天摄入维生素 D 补充剂存着重大争议，这与我的研究范畴相关。不光是我的亲朋好友经常问我，连我在世界各地参加科学会议，也会在我演讲结束后被问到。几十年前当我事业刚起步时，对于维生素 D 缺乏症的定义相对清晰，我们也知道此类病患为数不多。后来医学界根据血液中

measurement of a vitamin D compound in blood. Almost overnight, most of the world suddenly "became" vitamin D deficient or insufficient, just like that.

And, for those recently pushing for supplementing nearly everyone with vitamin D, the recommended dose suddenly became five to tenfold higher than before. So, you now have a sudden new dilemma, should you take vitamin D, especially in large doses, or not? The dogma in past decades is now defunct, and a new dogma is in place.

In fact, dogma change is relatively common, especially in clinical nutrition research. The dogma for years was that eggs and shrimp/lobsters were bad for blood cholesterol, and millions avoided eating these foods, only to find that this dogma is no longer accepted by "experts." Although, not everyone knows the reversal has occurred, so fear of eggs and shrimp/lobsters is still quite prevalent. Every decade or so, a new nutrient is either promoted, or maligned, and my instinct during these battles is now to just *wait 10 or 20 years before taking sides*! Even if the dogma is very dogmatic, and "experts" are fighting mad when challenged, ultimately, that dogma might change.

I love science and being a scientist. But, flip-flops in science bewilder many. I like to tell students in scientific research that we should always be open-eyed skeptics, challenging existing dogmas, which is how science "progresses." And, there are frauds and lies that creep into science, which get perpetuated for years, even though we forget, or even deny, them.

One myth that we scientists easily accept is the concept of "consensus." We say something like "*most scientists believe...*" as if that "solves the problem." However, voting is intrinsically "unscientific," because science is *not* really based on voted opinions. Truth is truth,

17. 做个怀疑者（上）：信条恒变
17. Be a Skeptic Part A: Dogmas Come and Go

维生素D化合物的含量，重新界定了"缺乏"和"不足"的定义。几乎一夜之间，世界上大多数人都"变成"了维生素D缺乏或不足者，就是这样而已。

更有甚者，近年有些人主张人人都要补充维生素D，他们所建议的剂量为以前的5到10倍。于是出现了一个新的难题：常人应该服用维生素D吗？尤其是需要大剂量补充吗？过去几十年的信条（共同信守的准则）被推翻了，有新的信条取而代之。

实际上，信条的改变是普通寻常的，尤其是在临床营养学的研究中。过去多年以来的信条认为蛋和虾/龙虾对血胆固醇不好，无数人因此戒绝此类食物，不料如今那些"专家"不再认同这个信条。当然，很多人还没意识到这种逆转，所以还是有不少人害怕吃蛋和虾/龙虾。每十年左右，就有一种新的营养或被推崇，或被推翻。在这一类争议中，我现在的本能反应是**先等个10至20年再来看哪方是对的**！尽管某个信条看来很有权威，尽管那些"专家"在遭到质疑时尽力奋起还击，但是最终那信条还是可能会改变。

我热爱科学，也热爱做科学家。但是，科学范畴中的翻来复去令很多人困惑。我告诫我的学生，在科学研究中，我们应该做个警惕的怀疑者，挑战现存的信条，这样科学才会"进步"。并且科学有时也会被弄虚作假者歪曲，尽管我们或会忘记甚至否定这种事，其影响却依然会延续经年。

regardless of the percent of "believers." The loaded word "believe" has implication of "faith", which we as scientists, often might deride, except when it is in "scientific consensus reports." This majority faith-based system, usually means we *trust* "leaders of the field" to "decide" on an issue. This is convenient, but also how dogma starts. The next thing that could happen is a possible "oppression" of the "minority" viewpoint, or those of a different "belief" system. I have sat on many illustrious "panels of experts," being asked to decide by voting what is the "right belief." The history of science, however, tells us, ironically, it is more often the "oppressed," the cynic, and the skeptic that is ultimately right, and *not* the "majority."

As an illustration, when I was a medical school student, decades ago, a central dogma of physiology was being taught, amply illustrated, indeed preached everywhere in the world, that "ontogeny recapitulates phylogeny." These words coined by Haeckel meant that as the embryo grows into a fetus, the process (*ontogeny*) follows the exact sequence of macro evolution, from fish to human (*phylogeny*). Thus, humans developed in the womb through a phase with gills, just like fish with gills are part of the phylogeny of evolution. How beautiful and clear were the steps. There were many vivid photographs of human embryos which absolutely illustrated and proved how macro evolution, represented by phylogeny, actually is "recapitulated" in utero.

Except, there are no gills in human embryos (the wrinkles just photograph like gills). *Except*, the photographs themselves in earlier textbooks were fraudulent (Haeckel made some of them up). And yet generations of students were taught, and believed, that these gills were *dogma*. Until the dogma blew up. But, how many still think the old dogma is true? At least one renowned professor of biology who was asked - a

17. 做个怀疑者（上）：信条恒变
17. Be a Skeptic Part A: Dogmas Come and Go

我们科学家很容易相信一个神话，就是"共识"这个概念。我们会说"**大多数科学家相信……**"，似乎这就能"解决问题"。但是，投票本来就是"不科学"的，因为科学**并不是**根据投票意见决定的。真理就是真理，不管有百分之几的人"相信"。"相信"这个词已经隐含了"信念"的意思，我们科学家常常对此嗤之以鼻，除非是写在"科学共识报告"当中。这种"基于大多数人的信念"的系统，一般意味着我们**信任**"这个范畴的领头人"对于某一议题作出的"判定"。这很方便，但是信条就是这样开始的。然后可能就会有人"打压""少数人"或者"异己者"的意见。我参与过很多赫赫有名的"专家小组"，被要求投票决定何为"正确的想法"。但是讽刺的是，科学史告诉我们，最终被证明是对的，往往是那些"被打压的"、持猜忌、怀疑态度的人，而**不是**"大多数"的人。

举个例说，几十年前我读医学院的时候，生理学上盛极一时、在全世界广为传颂的一个主要信条是"重演说"。海克尔发明的这个理论宣称，胚胎发育成胎儿的过程（**个体发生**），完全是按照从鱼类到人类的宏观演化（**种系发生**）的次序。因此，人类胎儿在子宫里发育，有一个阶段也有鳃，就像在物种演化过程中的有鳃的鱼类一样。海克尔把这个过程描述得美丽而清楚，并且发布了很多栩栩如生的人类胚胎照片，说明和证明宏观进化的过程（种系发生）如何在子宫中"重演"。

只是，人类胚胎并没有鳃（只不过是一些皱褶在照片中看似是鳃）。只是，早期教科书中的那些照片本身就是海克尔伪

man trained in developmental biology in the early 1980's - still thought it was so, even as I was writing this article. The history of this widespread erroneous belief persisted, because it was gently, unobtrusively, silently buried. Instead of fraud being declared boldly as example of scientific hubris and embarrassment, textbooks were just changed. If we are to move forward in truth, we've got to stop covering up dogma problems.

For nearly half a century, Piltdown man was hailed as one of the greatest discoveries proving that a transition was finally found, between ape and man. The exciting discovery produced accolades and awards for the discoverer, and educated millions of high school students worldwide about the certainty of apes becoming man. Through textbooks upon textbooks. *Until*, it was exploded as a total fraud, a combination of bones from different animals, chemically stained to look old. *Until* someone smart enough, and skeptical enough, discovered the fakery. In this case there was also a quiet burial, and little humility. My medical school days were 6 years after this discovery, but it was still in the books and there was not a word about this embarrassment.

Louis Leakey is considered by many as the greatest paleontologist of all time, discovering fossils which he claimed represented transition of apes to men. He was acclaimed worldwide, with findings widely taught. *Until* his wife Mary Leakey published her book *Disclosing the Past*, which provided fascinating evidence that many of the native workers, knowing *what the boss wanted*, provided him with, what in science we call "selectively biased samples;" and that much of the data were inconclusive. I even went to the museum in Kenya that celebrated Louis Leakey's discoveries, and was told, even though it was decades later, that the fossils were still in "safe custody under investigation," and were not for public viewing. Why would these fossils be under wraps after all these years?

17. 做个怀疑者（上）：信条恒变
17. Be a Skeptic Part A: Dogmas Come and Go

造骗人的。尽管如此，一代代的学生已经被这样教导，并且相信，这些鳃就是他们的**信条**。后来这个信条被推翻了，但是，究竟有多少人依然相信这个旧信条是真的？到我写这篇文章的一刻，至少还有一位大名鼎鼎的生物学教授依然相信（他是在1980年代初学习发育生物学的）。这个广为流传的谬误继续被传下来，因为这事情被人悄悄地、默默地掩藏了。教科书被修订了，但是没有人大胆地公开宣告这个骗局是科学界傲慢和尴尬的例子。如果我们想要在真理上进步，就不能一直掩饰信条的问题。

海克尔的"重演说：个体发生重演种系发生"已经土崩瓦解，华丽的辞藻会掩盖欺诈和伪论。要常存怀疑之心。
Haeckel's discredited theory of "recapitulation: ontogeny recapitulates phylogeny," fancy words covering fraud and misrepresentation. Be skeptical.

轰动一时的皮尔当人，精心塑造而成，曾广为传播，大获好评，直到 40 年后被揭穿。
The great Piltdown man, well stained, well disseminated, well respected, until exposure 40 years later.

From my personal research and academic background in clinical research, data are transparent, subject to double-checking by others; there is extensive emphasis on *observable* testing and retesting, statistics, population data, large randomized controlled trials, and challenges and challenges, especially of dogma. So, I find the field of *paleontology* to be extremely different. Of course, Leakey's findings had fallen under scientific disfavor apart from Mary Leakey's book, owing to other conflicting digs.

There is a famous drawing of apes to pre-man, to man, walking in a vivid line of progression that millions have seen in museums, school room walls, and textbooks, in color, in multiple languages. Very impressive. Indeed, most impressionable children in the world have seen this, and

17. 做个怀疑者（上）：信条恒变
17. Be a Skeptic Part A: Dogmas Come and Go

曾经有将近半个世纪，皮尔当人被誉为最伟大的发现，此发现终于证实了猿类进化到人类中间的过渡状态。这个振奋人心的发现令其发现者名成利就。借着一波又一波的教科书，全世界无数的高中生被教导：猿类变成人类是无可争议的。**直到有一天**，有人够聪明、够怀疑，才揭穿了这个骗局。原来皮尔当人根本是造假的，是用不同动物的骨头组合而成的，并且用了化学方法使它看起来年代久远。这件事也是被悄悄地掩藏了，没有公开的谴责。我读医学院，是在皮尔当人被揭穿六年之后，当时的教科书还在讲皮尔当人，并且完全没有提及这是一出荒唐的闹剧。

路易斯·李奇被许多人称为史上最伟大的古生物学家，他声称他所发现的一些化石代表着从猿类进化为人类的过渡状态。他享誉全球，他的学说被广为传讲。**直到有一天**，他的妻子玛丽·李奇出版了《往事揭秘》一书。书中提供了引人入胜的证据，证明有很多当地的发掘工人为**取悦他们的雇主**，给他找来的多是科学界所称的"有选择偏差的样本"，因此很多数据都是没有说服力、不能定论的。我甚至亲身去过肯尼亚那间表扬路易斯·李奇成果的博物馆，但是被告知，即使已经事隔几十年，那些化石仍在"安全保管调查中"，所以不能公开展出。为何过了这么多年，这些化石还是被包裹起来不让看？

从我在临床研究方面的个人研究及学术经验来看，数据应该是透明的，公开出来以供他人核实；科学研究特别强调**可观**

undoubtedly have "believed" this "scientific history" without a second thought. Thus, I appreciate the candor when some of these drawings, including at some Natural History museums, try to include, in the caption under the illustration, that this is an "artist's rendition" of the evolution of man. *Except* it is in the *fine print*, and maybe not everybody understands that "artist's rendition" is a charitable way to say "*imaginary*." Artists are much appreciated in communication of ideas, but art is not science nor data, yet this extraordinarily widely viewed painting has become itself cultural scientific dogma.

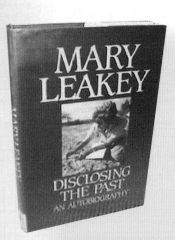

玛丽·李奇:"预先挑选数据。"

- 路易不在的時候,娣·辛普森負責運作,她會檢查每一堆化石,把路易可能會喜歡的某幾件挑選出來。
- 然後當路易來到加州,他會從那些已經挑出來留給他的化石當中再挑選某幾件,然後宣稱這些就是考古發現。
- 最後被精選出來的樣本只是所有化石中微乎其微的一部分。

"Pre-selection of data": Mary Leakey

- Dee Simpson, who directed operations in Louis's absence, then sorted through each heap and selected certain pieces as possibly pleasing to Louis.
- Later, whenever Louis happened to be in California next, he would take certain pieces from the selection kept for him and pronounced them to be artefacts.
- The proportion of poeces that finally made the grade was infinitesimal compared to the whole.

预先挑选数据以证明一个有偏差的结论。
Pre-selection of data to prove a bias.

17. 做个怀疑者（上）：信条恒变
17. Be a Skeptic Part A: Dogmas Come and Go

察的验证再验证、统计、总体数据、大规模随机对照试验，以及不断地怀疑与挑战，对于信条更应如此。我发现在**古生物学**范畴却截然不同。当然，随着玛丽·李奇的书出版，以及其他与之矛盾的考古发现，李奇的研究成果已经被科学界淘汰。

　　许多人都在博物馆、教室墙上和教科书里，看到过一幅用各种语言介绍的彩色名画，生动地描绘了从猿类迈向史前人类再到人类的进化过程，确实令人印象深刻。实际上，世界上有许多易受影响的孩子看过此画就毫不怀疑，理所当然地"相信"这是"有科学根据的历史"。因此，我很欣赏有些人（包括在一些自然历史博物馆里）坦白地在此画下方标上"此为艺术演绎"的注解。**可是**注解所用**字体很小**，而且未必人人都意识到"艺术演绎"只是"想像"的婉转说法。艺术家所注重的是传递想法，艺术不是科学或数据，然而这幅广为流传的画本身却已经成为文化科学的信条。

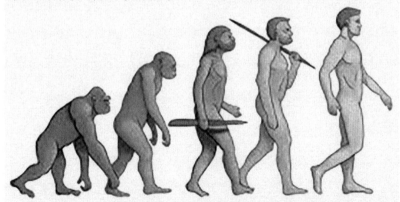

艺术效果非常好。但是未必有人留意到小字注解：艺术家的演绎。
Artistry is effective. Fine print might not be noticed: "Artists' rendition".

To my great surprise, I have seen reports of academic professors trying to prevent potential medical students from entering medical school, on the basis that "you can't be a medical doctor unless you *believe* in (macro-)evolution." What does belief in (macro-)evolution have to do with qualifications for a physician, who is supposed to be trained in solid verifiable science? The medical school candidate who can think for himself or herself about the complex unsolved mysteries of life, the skeptic and cynic, may be exactly the type of doctor we need.

Being a skeptic and cynic about dogmas is good.

Continued in Part B: Genes, Bones, and Emperors...

17. 做个怀疑者（上）：信条恒变
17. Be a Skeptic Part A: Dogmas Come and Go

让我很惊讶的是，我读到过一些报导说，有些教授试图阻止一些学生进医学院，理由是"除非你**相信**（宏观）演化，否则你无法成为医生"。对于（宏观）演化的认同与否，跟医生的资格有何相干？进医学院本来就是要接受扎实的实证科学训练。那些能够主动去思考生命复杂奥秘的准医学生，那些持怀疑和猜忌态度的人，或许正是我们所需要的那种医生。

对于信条常存怀疑和猜忌之心是好的。

未完待续：基因、骨头和皇帝……

翻译：杨迪霞

18. Be a Skeptic Part B: Genes, Bones, and Emperors

...Continued from Part A: Dogmas Come and Go

Soon after I started my residency training in pediatrics, a disease called sickle cell anemia was touted as the greatest example of human macro-evolution. You can see my description in Uncle Reggie stories, "Sickle Cell". Sickle-cell mutation was declared dogmatically as the major evolutionary "proof," as an "advantageous" mutation, because of its theoretic protection from malaria. My original skepticism of this concept was that the "protection" was based on "guilt by association", not by direct proof. And, now the concept is *collapsing*, since our country is spending millions of dollars trying to *eradicate* this genetic mutation, which to me, must prove that, by definition, it is *not* advantageous. This is a great development, and gives hope to many affected children. I have painful memories of managing many sickle cell children who experienced, repetitively, severe bone and chest pains from this potentially fatal disease. Clearly not an advantageous major mutation: I suspect that lab based philosophers had little idea of the real-life misery that their favorite mutation wrought on mankind, especially those of African origin. Be really skeptical when academicians are so dogmatic about anything.

In another Uncle Reggie story, I talked about the Shanghai cell, and

18. 做个怀疑者（下）：
基因、骨头和皇帝

……续上篇：信条恒变

在我的儿科住院医生训练刚开始不久，一种叫镰状红血球贫血症的疾病被吹捧为人类宏观演化最伟大的例证。你能在曾叔叔故事："镰状红血球"中看到我的描述。镰状红血球突变被当成信条般宣称为重大的进化"证据"，被形容为"有益的"突变，因为理论上它保护身体免于得疟疾。我对于这个观点所提出的怀疑是，这种"保护"是基于"关联的谬误"，而不是直接的证据。现在这个观点要**坍塌**了，因为美国正花费数以百万计美元试图**根除**这种基因突变，于我而言，这显然证明了这个突变**不是**有益的。这个医疗发展好得很，给许多受影响的小孩带来希望。我痛苦地记得我治疗过很多有镰状红血球的孩子，他们重复经历着由这种可能致命的疾病带来的严重骨骼和胸部疼痛。很明显这并**不是**一种有益的重大突变。我推测那些以实验为本的哲学家并不知道，他们最喜爱的突变在现实生活中带给人类（尤其是那些有非洲血统的）何等的痛苦。当学术界对

how *one cell* is more complicated than all of Shanghai. But during my residency days there were all these fantastic stories about how man could "finally" create life, so-called *spontaneously*, in a test tube, and everyone was ecstatic about a few amino acids that seemed to emerge. The hoopla was something like this, "we've done it, we don't need an Almighty Creator, we can do it ourselves, and surely, 'nature' can just *accidentally* do it!" However amino acids did *not* make life, no matter how dogmatic one tried to be about it. And, anyway if there is such a day when man can be so very smart that he can, "create" life, that still means that he used creative intelligence, and skilled hands to "create." It still means that there is no so-called "spontaneous," random, chance, accident-precipitated, "life." Regardless of how dogmatic this is expressed. And where did even the "simplest of cells," that is more complicated than London, New York, or indeed Shanghai come from? Each and every metropolis-like cell is a far, far, cry from any pile of dangling amino acids. Be skeptical.

There are all these comments in biology and in medical textbooks, dogmatically declared, that macro-evolution involves trials and errors, and chance, so it is inherently so *wasteful*, and there all these "vestigial things," such as organs left behind in the process, with no apparent redeeming value. See my Uncle Reggie stories again, to show that there really are *no vestigial organs left* to talk about. The only vestiges have been the theories, which have become even less than vestiges! Another dogma crash.

A variation of this vestigial organ concept, is the more modern so-called junk DNA, purported to be "vestigial left-overs" from previous evolutionary attempts at making DNA. At one time, junk DNA was declared dogmatically to be nearly 90% of the DNA, but the percent has drastically and steadily dropped, as there are more and more

18. 做个怀疑者（下）：基因、骨头和皇帝
18. Be a Skeptic Part B: Genes, Bones, and Emperors

基因治疗"治好"患血病的男孩，这种血病影响了数以百万计的人　　研究员朝着镰状红血球疾病的基因治疗迈进　　基因治疗显示有望对抗镰状红血球

> 重大突变促使物种进化的长期**最佳候选人**——著名的镰状红血球突变海报小孩。现在这个理论正在崩溃，因为美国投入数以亿计的美元试图消灭这个缺陷。
>
> The long lasting ***best candidate*** for advancing the species by major mutation, the famous sickle cell mutation poster child. Now the theory is crashing as the nation seeks to eliminate the defect by pouring in 100s of millions of dollars.

任何事情都奉为教条，我们真的要保持怀疑的态度。

在另一篇曾叔叔故事里，我谈到"上海细胞"，解释了一个细胞如何比整个上海更加复杂。但是在我担任住院医生期间，传来了一些荒诞的故事，关于人类"终于"在试管中创造出所谓自发的生命，人人都对几个似有若无的氨基酸欣喜若狂。他们如此大肆宣扬："我们做到了，我们不需要全能的造物主，我们自己就能做到。'自然界'肯定也能够**碰巧**做到！"但是，氨基酸**不能**制造生命，不管有些人如何把这个想法奉为教条。即使如果有那么一天，人类那么聪明能够"创造"生命，那就

所谓"垃圾DNA"像其他"残留概念"一样正在崩溃。那些发明"垃圾"说法的人应该把它当作垃圾扔掉。

So-called "junk DNA" crashing as another "vestigial concept." Those who coined this "junk" phrase could learn to eat the words. Like "junk food."

realizations that they are there for great purposes, sometimes even as the "orchestra *conductors*" of other DNAs, so technically they could be even more important, if you wish, than "the other" DNA. Someday, I like to predict, no junk DNA "vestige" will remain, as our understanding improves. As Ethel Waters, the ultra-famous singer once said, poignantly, "God don't make no junk." Another dogma crashing.

In trying to explain the complexity of his subject, a noted paleontologist, in a surprisingly candid commentary, stated that the "field of fossil rock study" was developing so fast, that, "in this field, you can turn over a rock in Africa, and re-write history." My cynical conclusion is that, if a scientific field can be changed so dramatically with just a new finding under a rock, it is likely immensely *unstable* and therefore by definition, *unreliable*. Meaning that today our dogmatic theories may reign in the field, but tomorrow it will be gone. Meaning that we *cannot "believe"* the theory today, since, who knows, it seems likely to be gone tomorrow. The changing theory is not reflecting how complicated a system it is, but likely how questionable our approach, or concepts, might be. In contrast, in fields of other sciences such as medical science, we investigate thousands of findings, subject them to open analysis, and *incrementally add* to the knowledge base, and

18. 做个怀疑者（下）：基因、骨头和皇帝
18. Be a Skeptic Part B: Genes, Bones, and Emperors

是指他运用富有创意的智力和熟练的手去"创造"。无论你表达得如何像教条一样，这仍然意味着没有所谓"自发的"、随机、偶然、碰巧而成的"生命"。那么，那个比伦敦、纽约或上海更加复杂的"最简单的细胞"从何而来？每一个都像大都市一样的细胞，与一堆毫无结构可言的氨基酸相比，差太远了。要常存怀疑之心。

在生物学和医学教科书中都有一些评论，教条式地宣称，宏观演化牵涉到反复的试验与错误，以及偶然，所以本来就是会如此**浪费**，会有这些"残留物"，例如在进化过程中剩下来的、看来不值得保留的器官。请再次读我的曾叔叔故事，你就会明白其实**已经没有残留器官剩下来**可以谈论了。只有那些理论才是残留的，后来更加连残留都没有！另一个教条崩溃了。

这种残留器官概念的一个变奏，就是更现代的所谓垃圾DNA，据称是从以前进化时尝试制造DNA而剩下来的"残留物"。起初有一段时间，他们教条式地宣布垃圾DNA占总数的近90%，但是这个百分比持续急剧下降，因为人们越来越意识到它们的存在有伟大的目的，有时甚至是担任其他DNA的"交响乐团**指挥**"，所以你可以说，在技术上它们可能甚至比"其他"DNA更加重要。我要预测，随着我们的知识进步，终有一天将再也没有垃圾DNA"残留物"的想法。正如超级著名的歌手埃塞尔·沃特斯的辛酸感言："上主不会制造垃圾。"又一个教条崩溃了。

there isn't this *instant discarding* of past findings and concepts with each new finding, under a rock. You might begin to see why, frankly, I'm very *cynical* about all of this.

Hans Christian Anderson's beloved story, translated into more than 100 languages, tells of the great Emperor who wants the most beautiful clothes possible. The cunning tailors play on his vanity, and that of the adoring crowds who fear pointing out the obvious. During the great parade, it is only the *innocent child* with "nothing to lose" who points out the basic problem. He is the ultimate skeptic and even cynic, who suddenly exposes the transparent truth.

One by one, dogmas crash and crash. Exposures begin, step by step. My advice to scientists, doctors, scholars and students, is to remain skeptical and even cynical. The Spontaneous Macro-Evolution Emperor has *fewer and fewer clothes*, maybe even no clothes. Keep asking questions, like the innocent child in the story, and keep challenging the system. Frankly, you can, and should do that with faith issues also. It's okay to challenge Christian history, principles and conduct, because that's how we can all learn. Even challenge the Scriptures to find out the truth. And no one will threaten you for making the challenge. The history of Christianity is full of attacks by skeptics and cynics over several thousand years. But the amazing thing is that history tells us that these attacks by skeptics and cynics, have often resulted dramatically in changed lives in the challengers themselves. Their skepticism often is finally answered, and they often become, surprisingly, strong defenders of the faith they once challenged. Be a skeptic and cynic, I urge you, and *challenge you*.

18. 做个怀疑者（下）：基因、骨头和皇帝
18. Be a Skeptic Part B: Genes, Bones, and Emperors

一位著名的古生物学家为了解释他的研究课题有多复杂，在一篇坦率得令人惊讶的评论中指出，"化石岩研究范畴"发展得如此之快，以至于"在这个范畴里，你在非洲翻转一块岩石，就可以重写历史"。我充满猜忌的结论是：如果某个科学范畴只因一块岩石下的一个新发现就可以发生剧变，那么它很可能极之**不稳定**，因此显然就是**不可靠**。这意味着今天我们的教条式理论在这个范畴称王称霸，但是明天它可能就消失了。又意味着我们**不能"相信"**今天的理论，因为它似乎很可能明天就消失，谁知道呢。不断变化的理论并非反映那个系统多么复杂，而是反映我们的方法或概念可能大有问题。相比之下，在诸如医学科学的其他科学范畴里，我们审查成千上万的发现，让它们接受公开的分析，并**逐步加添**到知识库中；我们不会每次在岩石下有新发现就**立即扔掉**过去的成果和观念。坦白说，我对这一切非常猜忌，你可能也开始明白为什么。

安徒生的一个深受爱戴的故事，被翻译成超过100种语言，讲述伟大的皇帝想要全世界最美丽的衣服。狡猾的裁缝玩弄着皇帝的虚荣心，又知道敬爱他的群众不敢指出那显而易见的事实。在那场大型巡遊中，只有**天真的孩子**"无所顾忌"，指出了问题所在。那孩子是终极的怀疑者，甚至是猜忌者，突然揭穿了那透明的真相。

信条一步接一步被揭穿，一个接一个崩溃。我对科学家、医生、学者和学生的建议是保持怀疑甚至猜忌的态度。这位名为"自发的宏观演化"的皇帝所穿的**衣服越来越少**，甚至可能

Astronomer by Night, Canon by Day

When Nicolaus Copernicus wasn't redrawing the celestial map, he held down a day job as a Catholic canon (ecclesiastical administrator). As the Reformation grew rapidly and extended its influence in Poland, Copernicus and his respected friend Tiedemann Giese, later bishop of Varmia, remained open to some of the new ideas. Copernicus did not leave a written record of his views,

与一般神话相反，**敬畏上主的**哥白尼挑战以往大多数人的**科学**信条，提出太阳是我们"太阳系"的中心。正如图中文字"晚上是天文学家，日间是大教堂司铎"所示，他每天的工作是忠心地服务教会，同时也是个怀疑者。

Contrary to common mythology, *God-fearing* Copernicus challenged previous *scientific* majority dogma to propose that the sun is our center of our "solar system." As the caption Astronomer by Night, Canon by Day shows, his day job was serving the Church, which he did faithfully, even as a skeptic.

18. 做个怀疑者（下）：基因、骨头和皇帝
18. Be a Skeptic Part B: Genes, Bones, and Emperors

一丝不挂了。我们要像故事中天真的孩子一样，不断提问，不断挑战系统。坦白说，对待信仰的议题，你也可以、也应该这样做。你可以挑战基督信仰的历史、原则和行为，甚至挑战圣经经文、找出真相，因为我们全都是这样学习的。没有人会因为你作出挑战而威吓你。数千年来，基督信仰的历史中充满了怀疑者和猜忌者的攻击。但令人惊奇的是，历史告诉我们，怀疑者和猜忌者的这些攻击和挑战，通常会导致他们自己的生命发生巨大的改变。他们的怀疑通常到最终都得到解答，而且他们通常会变成强大的守护者，去捍卫他们曾经挑战过的那信仰，意想不到吧。我敦促你，并**挑战**你，做个怀疑者、猜忌者。

翻译：Hongyan Zhu

19. What's the Red Cross?

I've asked many people about what they know about the Red Cross, and the usual response is "it's that ambulance that comes to save people". There's a lot more to that story, I have found.

When I started my medical mission 20 years ago in China, I noticed that all the medical clinics and hospitals had the Red Cross sign very prominently displayed. So I asked a local doctor "is this related to the Christian faith?" He looked quite surprised, and he said "no, of course not; it just means a medical facility." In fact I began to realize that there were so many crosses in China, all over the place, because any medical facility or any ambulance basically had a Red Cross, and nobody really knew of any connection with the Christian cross. One hospital we visited in a small town in Yunnan actually had five crosses, including a huge one over the entrance of the hospital, although it wasn't colored red in that case.

One year I had the opportunity to visit my old ancestral village, where there was the only government hospital, clearly with a Red Cross over the entrance. I realized that there was even more to the story: the hospital originally was a mission hospital, where my grandfather was the hospital director. In essence, it was basically a Christian hospital, and so the cross was really totally appropriate.

During World War II, my father was a newly graduated doctor, and he told me about the world wide convention so that the hospital he worked in had a Red Cross painted on the rooftop. Aircraft from either

19. 何谓红十字？

我问过很多人对红十字有什么认识，最普遍的回答是："就是赶来救人的救护车。"我发现它有更多的涵义。

二十年前，当我在中国开始医疗服务的时候，我注意到所有诊所和医院都有非常显眼的红十字标志。我问一位当地医生："这个与基督信仰有关吗？"他一脸错愕，说："不，当然没有关系，这只是代表医疗机构。"事实上，我开始注意到在中国有非常多十字标志，到处都有，因为任何医疗机构或救护车基本上都有一个红十字。但是没有人真正知道它与基督信仰的十字架的关系。我们在云南小镇参观过的一间医院，居然有五个十字标志，其中一个大大的矗立在医院大门口之上，虽然这次它没有漆成红色。

某一年我有机会去拜访我祖上的村庄，那里只有一家公立医院，其大门口上方有一个显眼的红十字。我发现这故事不是那么简单：这医院起初是传教士所建的，我的祖父当时是这医院的院长。它的本质是一家基督教医院，所以十字标志是恰如其分的。

二战期间，我父亲是刚毕业不久的医生。他告诉我当时有

side would not bomb hospitals with a cross on the roof, knowing that they were medical facilities and were caring for patients. I was born in such a facility (the Queen Mary hospital of Hong Kong) since my father was on staff there. I'm glad there was such a convention. However in today's wars, especially in the Middle East, there's not that much respect to traditional conventions and we see Red Cross workers captured and even beheaded.

Some people have asked me "is it really a connection between the Red Cross and the Christian cross." Actually there is a particularly interesting story. Two Swiss men first started the Red Cross organization, to provide help to wounded soldiers on the battlefield, which then expanded to helping distressed people in need of medical care all over the world, a very remarkable goal. The Swiss flag itself is a white cross on a red background, representing the Christian heritage of the country. So they came up with the bright idea to invert the colors, to make the Red Cross on a white background, as a symbol of helping the wounded in need.

After I realized the connection, whenever I went to Western Europe or the United Kingdom, I started looking carefully at their flags, and realized that there is nearly always a symbol of the cross. In the UK Union Jack, there are actually three crosses: the cross of St. George for England, a red cross; the cross of St. Andrew for Scotland, a diagonal white cross; the cross of St. Patrick for Ireland (now Northern), a diagonal red cross. All Western European countries basically have some sort of Christian background, and this is a reminder of that heritage, which seems to have been quite neglected or forgotten in today's world.

In Asia, I have flown into South Korea many times, and every time as I was landing in one of the major cities, one of the most startling views to me was to see many crosses on top of churches over the entire city. I

19. 何谓红十字?
19. What's the Red Cross?

在中国,我祖父的医院上面的红十字。
Red Cross over my grandfather's hospital in China

一个国际惯例,就是在医院的屋顶漆上红十字。交战双方的战机看见红十字,知道那是医疗机构,正在救治病人,就不会轰炸那个地方。我正是在这样的一家医院(香港玛丽医院)出生的,因为我父亲恰好供职于此。我很庆幸有这样的国际公约。但是现在的战争(尤其是在中东)不再那么尊重传统公约,我们看到红十字会的工作人员被俘虏,甚至被斩首。

有些人问我"红十字和基督信仰的十字架是不是真的有关系"。这里有一个相当有趣的故事。两个瑞士人创办了红十字会,帮助战场上的伤兵,后来扩展至帮助全世界需要医疗救助的落难人群,这真是了不起的目标。瑞士国旗是红底白十字,代表他们国家基督信仰的传承。他们灵机一动,转换颜色,以白底红十字作为帮助苦难者的标志。

韩国教堂顶的十字架。试想像城里所有十字架的霓虹灯全都点亮,而实况就是这样。

Cross over a Korean church: imagine if all the cross neon lights are on in a city, which is what actually happens.

realized that this had to be a coordinated effort, and a reflection of the modern day heavy Christian influence now in the country, estimated to be 40% Christian, one of the highest rates in Asia.

However, in going to the Middle East, I found the opposite, that even the Red Cross in the Red Cross organization was not appreciated, and most of the countries had decided to change the symbol to a Red Crescent; even though it is the same organization. Symbols do matter, and some areas of the world are more sensitive than others.

In fact indeed, the implications of the Red Cross are much more wonderful than we might assume. The main implication is that the Red Cross actually represents the cross of Christ, which to all Christians is a symbol of salvation and hope, because of the core belief that Christ died on the cross to save the world.

在韩国,这是教堂还是医院?(答案:这不是聚会的地方。)看颜色没有帮助。

So what is this in Korea, a church or a clinic? (Answer: it is not an ecclesia.) The color doesn't help.

19. 何谓红十字？
19. What's the Red Cross?

自从我了解到这个关系，每次我去西欧或者英国，都会仔细察看他们的国旗，总是能找到十字标志。英国国旗上有三个十字：红色的圣乔治十字代表英格兰；斜行白色的圣安德鲁十字，代表苏格兰；斜行红色的圣帕特里克十字，代表爱尔兰（现在叫北爱尔兰）。所有西欧国家基本上都带有某种程度的基督信仰背景，国旗上的十字图案正是提醒他们这种传承，可惜现今世界似乎忽略了或者已经忘记了。

在亚洲，我曾经多次飞到南韩，每次降落在某个大城市，看见无数教堂顶的十字架遍布全城，都让我感到非常震撼。我意识到这一定是协调努力的成果，同时反映目前基督信仰对这个国家的现代社会影响深远，估计基督徒占总人口40%，在亚洲算是很高的比例。

但是在中东恰恰相反，连红十字会的红十字也不为人所接受，很多国家决定将其标志改为红新月，尽管是同一组织。标志是重要的，而世界上某些地区比其他地区更敏感。

事实上，红十字的涵义比我们所想像的要精彩得多。红十字其实主要代表基督的十字架。对于所有基督徒而言，它是拯救和盼望的象征，因为基督信仰的核心是基督为拯救世人而死在十字架上。

翻译：Sonic

20. Personal Testimony: A Clinician Scientist Perspective

Sometimes I hear a question, "you are a scientist, so why do you believe in God?" Actually, my answer is, something like this, "it's *because* I'm a scientist, and I see the amazing complexity in the human body, that I am driven to an *overwhelming conclusion*, that, logically, there must be a super, super, intelligence that has been able to design such a wonderful body. And I cannot think of any other way that it could have happened." Certainly, I just cannot imagine "how" it could happen through millions of little changes that happen through millions of years. And I know, that, *regardless of how one views how we got here*, we have the reality of how to live out our life with *meaning*, versus those who feel there is no meaning in life, and therein I think lies the real story.

In fact, I have been a *very serious* scientist.[1] No kidding. Some people like to Google me, as Reginald Tsang (surprisingly, there is only *one* such name), to have some fun checking me out. I certainly was privileged to enjoy a lot of personal fun in the colorful research and clinical academic world, but through it all, my faith in a Creator God only increased with time.

---

1. Reginald Tsang MD, was principal investigator for NIH (National Institutes of Health) grants that totaled over $30 million, and published more than 400 papers. He was formerly chief of one of the top Neonatology Divisions in USA, and vice Chairman of the Pediatrics Department, Cincinnati Children's Hospital Medical Center.

20. 个人的见证：
临床医生科学家的观点

有时我会被问到："你是科学家，怎么会信上主？"实际上，我的回答大概是这样的："正因为我是科学家，我看到人体复杂得让人难以置信，让我得出如下**压倒性的结论**：根据逻辑，如此奇妙的身体，必定是由一位超级的超级智慧设计出来的。除此之外我想不出还有其他的方法。"当然，我根本想像不出数百万年间发生数百万次细微改变"怎样"可以达成这个结果。但是我知道，**不论我们认为我们是如何出现的**，我们实际要面对的问题是如何活出生命的**意**义，而有些人却觉得生命毫无意义，我想真正的故事就在其间。

事实上，我是一个**非常严谨的科学家**。[1] 说真的。如果你有兴趣了解我多一点，可以在谷歌上搜寻 Reginald Tsang（想不到，这个名字仅此一个）。我的确是有幸，能够在多彩多姿的研究和临床学术世界中享受许多乐趣，但是通过这些经历，我对造物主的信仰却与日俱增。

---

1. 曾振锚医生曾经是美国国家卫生研究院（NIH）的首席研究员，其项目获资助总额超过三千万美元，并曾发表400多篇论文。他过往是美国顶尖的新生儿科领袖之一，曾担任辛辛那提儿童医院医学中心的儿科系副主席。

生命初始，步步奇迹。
Life begins, every step is a miracle.

I think that, because I was actively involved in neonatal care of premature infants, even infants below 500 grams birth weight, I had the special opportunity and privilege to also **connect** the theories of biology with the actual reality of life, and the connection "clinched the deal." Definitely I did not live my academic life just in theory, facing a computer or working in a rarefied lab, but I actually witnessed vividly the remarkable complexity being played out in real life, around the beginning of life.

I was heavily involved, as Director and founder of one of the first Perinatal Research Institutes in the country, in the research and clinical world of "perinatology," which is the interplay of mother, placenta, fetus and newborn life. In fact, in today's world, the avant-garde discipline of even *fetal surgery* is now a major component of perinatal science. This newer field implies that we appreciate that *fetal life* is an important life, and we can directly treat babies even before birth, even with surgery! This will sound amazing and even shocking to many, who might not even realize this surgical discipline exists.

20. 个人的见证：临床医生科学家的观点
20. Personal Testimony: A Clinician Scientist Perspective

我想，正因为我经常参与早产新生儿（有些出生体重甚至低于 500 克）的治疗，才特别有机会和荣幸去把生物学理论和现实的生命**联系**在一起，而这种联系"一拍即合"。毫无疑问，我不是活在学术象牙塔里，整天面对电脑或忙碌于净化的实验室中，我确实历历在目地见证过真实生命从其起始就展现出令人惊讶的复杂。

作为美国最早的围产期研究所之一的创办人及总监，我曾经密切参与"围产期学"的研究及临床工作。围产期学研究母亲、胎盘、胎儿及新生儿之间的相互作用。实际上，到了今时今日，就连前卫如**胎儿外科手术**也是围产期科学重要的一环。这个新的领域表明我们看重**胎儿的生命**，视之为重要的生命，我们甚至可以通过手术直接治疗出生前的胎儿！对于许多人而言这听起来很神奇，甚至是震惊，他们甚至不知道这种手术存在。

回溯到数月前，刚开始的时候，神奇的"第一个胚胎细胞"精准地分裂为不同的器官组织。早期的细胞在微观的层面精准地迁移，精准地迁移到正确的细微位置，精准地发育为每一个非常不同的复杂器官。我在各种重要的研究计划中担任指导者，与一丝不苟的科学家一起工作，感到十分兴奋。他们拼命地、煞费苦心地试图对影响这复杂生命工程不同阶段的每一种化学物进行解码。这就像成千上万的科学家试图合作拼凑一幅**巨大得不可思议的拼图遊戏**，这工程可远比成语故事中的盲人摸象困难得多。

And then backing up in time for a few months to the beginning, to that magic "first baby cell," dividing so precisely into different organ tissues, with early cells traveling so precisely at the micro level, precisely to the right micro location, precisely developing each complex yet so very different organ. I experienced the great excitement, in my role as director of key research programs, of working with meticulous scientists, who were desperately and painstakingly trying to decipher and decode each chemical that affects each step of this complex equation of life. It is all literally like working on an impossible *gigantic jigsaw puzzle*, which hundreds of thousands of scientists have been trying to piece together. Comparatively, the proverbial four blind men around the elephant had a totally easy job.

I'm excited especially about birth itself, and truly feel that *every birth is a miracle*! From the amazing placenta that has been transferring each and every necessary *nutrient* from mother to fetal baby, while disposing of all individual items of *rubbish* that the fetal baby generates, back to the mother. And how the moment of birth precipitates thousands of changes that happen in the baby within seconds and minutes, in order to survive and live in a totally new environment, in air versus liquid, exposed versus protected, cold reality versus warm mother's insides. Everything has to work, just like that, click, click, click, just like that, and be perfect. It is an understatement to say it works like clockwork: it really works superbly better than thousands of the best Swiss watches.

And especially I enjoy the "magic switch," the dramatic changes of the baby's circulation and heart, from a fetal system, suddenly within seconds and minutes, into the neonatal and adult system. Practically *everything* in the circulation and heart has to change. Who designed all the *switches*, who designed the changes in *directions* of blood flow, who

20. 个人的见证：临床医生科学家的观点
20. Personal Testimony: A Clinician Scientist Perspective

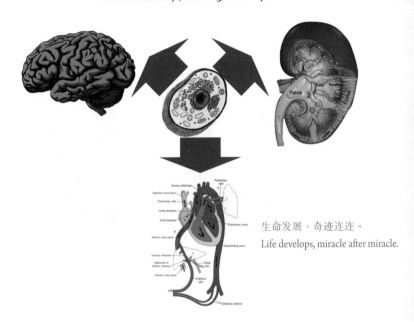

生命发展，奇迹连连。
Life develops, miracle after miracle.

我对出生这件事特别感到兴奋，真真切切觉得**每一次出生都是奇迹**！首先是神奇的胎盘，它从母体把每一样必需的**营养**传输到胎儿，同时把胎儿所产生的每一样**废物**运送回母体。然后在出生那一刻的分秒间，从水里到空气中，从被保护到暴露，从妈妈温暖的体内到冷冰冰的环境中，为了在一个完全陌生的新环境存活下来，婴儿的体内瞬间发生了成千上万的变化。每一件器官组织都要运作起来，咔哒、咔哒、咔哒，就像这样精准无误。以时钟工艺作比喻也不足以形容，婴儿的出生实在比上千枚最高级的瑞士手表还要细致精确得多。

我尤其惊叹那"魔法转换"，让婴儿的心脏和循环系统在分秒间魔法般改变，骤然从胎儿系统转换为新生儿和成人的系统。心脏和循环系统中几乎**一切**都要改变。到底是谁设计了这

designed the *opening and closing* of holes and channels? It is literally mind-boggling to even think about it.

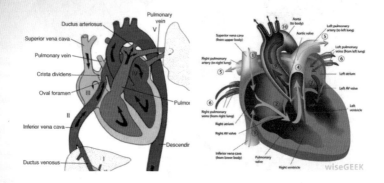

胎儿（左图）与成人（右图）的心脏和循环系统有显著分别。
The huge difference between fetal (left) and adult (right) heart and circulation.

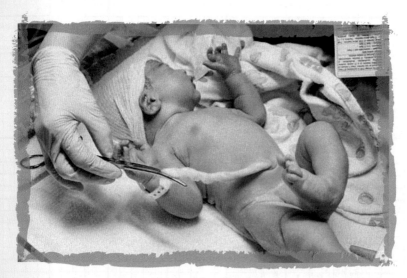

系统转换的关键时刻。
The critical moment of transition.

20. 个人的见证：临床医生科学家的观点
20. Personal Testimony: A Clinician Scientist Perspective

些转换，设计了血流方向的改变，设计了腔隙和瓣膜的开合？单是想想就已经觉得不可思议。

翻译：Sonic

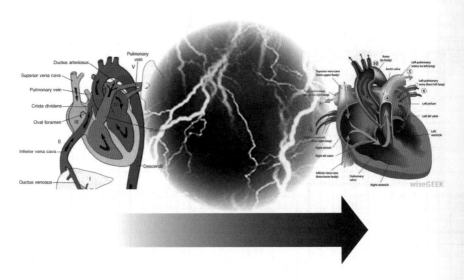

"魔法转换"：在关键瞬间，血流方向从胎儿的心脏和循环系统无缝转换为成人系统。

The "magic switch" from fetal to adult heart and circulation, in vital seconds and minutes, blood flows and directions switching on cue seamlessly.

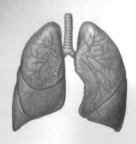
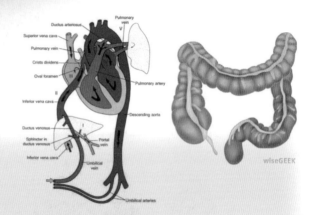

一个胎盘细胞　　One Placental Cell

- 肠：食物
- 肺：氧和二氧化碳
- 肾：废物
- 肝：制造蛋白质
- 内分泌：讯号
- 不能尽录

- Intestine: food
- Lungs: O2 and CO2
- Kidney: wastes
- Liver: makes proteins
- Endocrine: signals
- Etc.

巨大的胎盘细胞有神奇的特质和功能，让生命在出生前就受到爱的滋养。

Life nurtured lovingly even before birth, by essentially one huge placental cell with magical properties and functions.

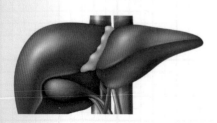
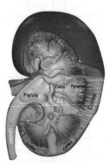

Postscript

The amazing miracle of birth is one big reason why in my own research career, I was fascinated with studying this transition phase, because essentially *everything* has to change, and change is always interesting and potentially exciting to study "how" it all works. "How" is truly the province of the scientist, and I was most pleased that the Creator created such a complex system that I could study, apply for grants, and write papers about events at this exciting time. How? How? How? Each "how" is a challenge for scientists to try to solve each mystery.

Sometimes I even joke that we should truly thank the Designer for all of this complexity, otherwise there would not be hundreds of thousands of scholars that come from Asia and the rest of the world to study in the USA, so that we could meet many as great friends!

I would think that there is no one in biologic sciences who would deny that the complexity and precision that we see in the body is definitely breath-taking. There's just no way, I feel, that we can brush it off as "O, it just happened, with no real design, and totally by *accident*." Or "we are here as a result of zillions of accidents." We are just an accident? Or zillions of accidents? I find that even more difficult to understand. I often like to teach that what we see is intelligent design, infinite complexity, invisible complexity, instinctive complexity, and impossible mathematical complexity, which altogether can only mean a *super intelligent designer*

后记

诞生的奇迹令我在研究生涯中非常醉心于研究这个转换过程，因为基本上**一切**都要改变，而改变总是让人很有兴趣，也可能会觉得刺激，想去探究它是"怎样发生"的。"怎样发生"的确是科学家关注的重点。我很高兴造物主创造了这样复杂的系统，让我可以进行研究、申请资助、撰写论文，探讨这个令人兴奋的时刻所发生的事件。怎样？怎样？怎样？每一个"怎样"都挑战着科学家去尝试解开所有奥秘。

有时我甚至开玩笑说，我们真的应该感谢造物主将一切都设计成如此复杂，否则就不会有成千上万的学者从亚洲和世界各地来到美国进行研究，我们就不能结交这么多好朋友！

我想没有任何生物科学家会否认，我们所看到的身体复杂而精巧得令人叹为观止。我认为我们不可能只是轻描淡写地说一声"哦，这些都只是**意外**，没有谁精心设计"或者"我们之所以在这里全是因为无数的意外"。我们只是一场意外吗？还是无数次的意外？我觉得这个说法更难理解。我在教学的时候经常说，我们所看到的是满有智慧的设计，无限的复杂、看不见的复杂、天生的复杂、难以计量的复杂，这一切只能归结于

who designed all of this.

And it is not like the Creator has not given us evidences for His existence. He gives us stunning evidences, many "signatures" of his design and creation. Just *open our eyes*, the evidences are staring at us right in the face. And these signatures are also artistically "beautiful" to behold, not just a boring "thing." Especially for medical scientists and medical doctors. There is just "no excuse," in my view, to say that there is no "evidence." Today, all of us know infinitely more about everything in our intricately designed body, than philosophers and scientists of even 150 years ago, so I would dare say we truly have even more reasons to believe.

Actually, we also even have the role models of thousands of dedicated scientists who created the modern western revolution in science that we read about in our textbooks. What is often not told us is that these great scientists basically assumed that there was a Creator, a Grand Designer, and that often they felt that they were just part of the great discovery process of finding out what the Creator was all about. I like to quote the very famous Asian philosopher-apologist Dr. Stephen Tong, who declared, and I paraphrase, "science is basically *created* man, using *created* reasoning (the brain), studying *creation*, trying to understand or *discover* (which means take off the cover) the principles that *God has created*."

The greatest astronomer Johannes Kepler, dramatically stated that his personal drive in studying astronomy was trying to "think God's thoughts after Him." In all humility, we will admit that, in the natural sciences, we are not creating anything in our own research, but only really just discovering *what is already there*, which is amazing enough, and maybe at best making some *rearrangements* of the basic components we find during our discoveries.

后记 Postscript

一位**拥有超级智慧的设计师**。

造物主并不是没有把他存在的证据给我们看。在他的设计和创造中有许多"签名",这些都是令人目眩的证据。只要**睁开双眼**,证据就在我们面前。这些签名作品并不是普普通通的"东西",而是呈现着"艺术之美",对于医学科学家和医生而言尤其如是。在我的眼里,我们"没有借口"说看不到"证据"。今天,我们任何人对于设计得复杂无比的身体所知道的,都远远超过 150 年前的哲学家和科学家,因此我敢说我们实在更加有理由去相信这位造物主。

事实上,我们还有数千位专心致志的科学家作我们的榜样,他们开创了我们在教科书上所读到的现代西方科学革命。只是他们通常没有告诉我们,这些伟大的科学家基本上都认为有一位造物主,一位伟大的设计师,而他们常常感到他们只是在探索并认识这位造物主的过程中的参与者。我喜欢引用大名鼎鼎的亚洲哲学辩论家唐崇荣博士的观点,大概的意思是:"科学,基本上就是**被创造**的人运用**被创造**的思考方法(脑子),去研究**被创造**的一切,试图理解或**揭露**(就像掀开盖子)**上主所创造**的那些原则。"

伟大的天文学家约翰内斯 · 开普勒戏剧性地声称,他研究天文学的动力就是试图"追随着上主的想法来思想"。我们都得谦虚地承认,在自然科学界,在我们自己的研究里,我们什么也没有创造过。仅仅是探索那些**已经存在的**,已经足以令我

科学与创造如何连系
How science and creation are linked

科学与创造

- 被创造的人运用
- 被创造的思考方法（脑子）
- 研究被创造的一切
- 去理解或揭露（"掀开"盖子）上主所创造的原则

(改编自"亚洲的葛培理"唐崇荣博士)

Science and Creation

- Created man using
- Created reasoning (brain)
- Studies Creation
- To understand, discover ("take off" cover) principles God Created

(Adapted from "Billy Graham of Asia" Stephen Tong)

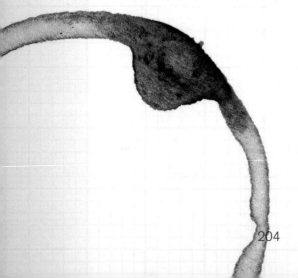

后记
Postscript

们啧啧称奇。或者我们顶多把探索时找到的那些基本构件重新整理一番。

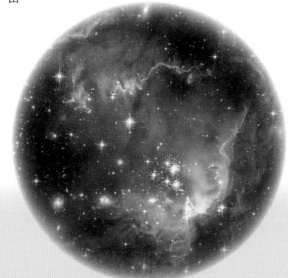

翻译：Sonic

所有逻辑中的逻辑
The logic of all logic

满有智慧的设计	Intelligent Design
无限的复杂	Infinite Complexity
看不见的复杂	Invisible Complexity
天生的复杂	Instinctive Complexity
难以计量的复杂	Impossible Mathematical Complexity
= 拥有超级智慧的设计师	= Super Intelligent Designer

About the Author

The author Reginald Tsang, is a medical doctor who specializes in premature infant care (Neonatology). He was Director of the Division of Neonatology at he Cincinnati Children's Hospital Medical Center for 15 years. He has published more than 400 scientific articles and papers related especially to infant and perinatal calcium and nutrition research. In 1994 he took early retirement to answer the call for medical missions, co-founding the Medical Services International organization to work in rural minority areas in Southwest China. In 2004 he took his second retirement to serve his home church, the Cincinnati Chinese Church where he is founding elder, serving especially in youth ministry. Professor Tsang is affectionately called "Uncle Reggie" by 3 year old kids to 70 year old adults, since he loves to chat and tell many stories related to his many travels overseas. He has logged in 3 million miles of flight travel, taught 10,000 pupil hours of English especially to Chinese village kids, and helped bring more than 5,000 short term people units to China and Southeast Asia.

Professional recognitions include: President American College of Nutrition; President Midwest Society of Pediatrics; American Academy of Pediatrics Nutrition Award; Bristol Myers Nutrition Award; Cincinnati Children's Hospital Founding Executive Director of the Perinatal Research Institute; University of Cincinnati Medical School's highest academic award, the Drake award. In all these instances he was the first Asian or Chinese American to be so recognized.

作者简介

作者曾振锚是一位专门从事早产儿护理（新生儿科）的医生。他曾在辛辛那提儿童医院医学中心担任新生儿科主任15年。他发表了400多篇科学文章和研究论文，特别专注于婴儿及围产期钙质和营养的研究。1994年，他回应医疗服务使命的呼召，提早退休，与伙伴共同成立了国际医疗服务机构，服事中国西南地区少数民族的农村。2004年，他第二次退休，回到他的母会辛辛那提华人教会（他也是该教会的创会长老），致力参与青年事工。从3岁小孩到70岁长者都亲切地称呼曾教授为"曾叔叔"，因为他喜欢聊天，又会讲很多他海外旅行的故事。他的飞行里数已达300万英里，教授过10,000教学时数的英语，特别是教导中国乡村的小孩子，又有份带领5,000多人次到中国和东南亚参与短期海外工作。

专业认可资历：美国营养学院院长；中西部儿科学会主席；美国儿科学会营养学奖；Bristol Myers营养学奖；辛辛那提儿童医院围产期研究所创办人及执行总监；辛辛那提大学医学院最高学术奖——德雷克奖。整体来说，他是第一位得到如此认可资历的亚裔或华裔美国人。

Books published by Dr. Reginald Tsang
曾振锚医生著作

1. Coffee with Uncle Reggie
2. 与曾叔叔闲聊
3. Starting an Academic Medical Career
4. 医学学术成功起步
5. Nutritional Needs of the Preterm Infant
6. 早产儿营养需要
7. Nutrition During Infancy: Principles and Practice
8. 婴儿营养原理与实践
9. Calcium Nutriture in Mothers and Children
10. Nutrition and Bone Development (with JP Bonjour)
11. Nutrition During Infancy
12. Nutrition in Preterm Infants: Scientific Basis and Practical Guidelines
13. Mineral Requirements of Preterm Infants
14. Textbook of Neonatal Medicine: A Chinese Perspective (with Chief Editor Victor Yu)

The author invites you to read more Uncle Reggie stories at his bilingual website: Reggietales.org.

Publishing and printing of this book is supported by YFAN Heritage Foundation. You can support this effort through USA tax deductible donations to:
YFAN heritage foundation, c/o 1002 Eastgate Dr., Cincinnati, OH 45231
Check made out to: YFAN heritage foundation, Memo line: YFAN literature mission fund

作者诚邀你登上他的双语网站，阅读更多曾叔叔的故事：Reggietales.org。

本书的出版和印刷获得 YFAN Heritage Foundation 赞助。你可以捐款支持。捐款收据可申请减免美国税项。支票请邮寄至：
YFAN heritage foundation, c/o 1002 Eastgate Dr., Cincinnati, OH 45231, USA
支票抬头：YFAN heritage foundation；备注（如适用）：YFAN literature mission fund